"A fascinating and deep
discovery and innovat
and utilization of autc
sightful and compellir.

MW01484091

trajectory of the science, but a sense of how valuable robotics technologies will be in addressing society's most critical problems. A 'must-read' work for current and future generations of ocean scientists, technologists, engineers, and policy makers."

—Richard Spinrad, PhD; Former Chief Scientist and Under Secretary
of Commerce for Oceans and Atmosphere, NOAA;
Professor Emeritus of Oceanography, Oregon State University

"The technical complexities and geopolitical issues surrounding exploration of the sea mirror those for space, yet we know much more about the surface of the Moon and even Mars than we do about the chasms and mountain ranges of Earth's sea floors. Bellingham's book, which illuminates the desires, challenges, innovations, and rewards of exploration in extreme ocean conditions, stirs the same emotions and curiosity that led me toward venturing into space. I'm grateful for these scientists, engineers, and inventors who look down far below the stars and reach into the most inhospitable places here on Earth to reveal its wonders and possibilities."

—Daniel Tani, NASA astronaut, space shuttles *Endeavor* and
Discovery; NASA aquanaut, NEEMO 2; Director of Human
Exploration Operations, Northrop Grumman

"In the many volumes on life in the deep sea that line my bookcases, the equipment used to explore the depths is generally mentioned *en passant*. Dr. Bellingham's book is quite different, and quite welcome, for it describes in considerable detail the physical conditions of the ocean's terrain and the complex array of equipment used— yesterday and today—to explore it. An interesting and readable account of the history and potential of marine exploration technology—at last!"

—David L. Pawson, PhD; global deep sea submersible explorer;
Senior Scientist Emeritus and former Director,
National Museum of Natural History, Smithsonian Institution

"This book reads like an adventure novel of explorations from the Mariana Trench to the polar seas, told through marine inventions from early submarines to today's remotely operated vehicles that can operate for months without human intervention. At each turn Bellingham and Geib present real-world events to frame the impacts of these complex and elegant systems here on Earth, and how they might one day search for life in the seas of our neighboring planets and moons."

—Rear Admiral Lorin Selby, US Navy (Ret); former Chief of Naval Research

"An engaging account of vital marine explorations and innovations. Many challenges remain in the quest to understand the physical and biological systems of the oceans well enough to protect them even as we humans expand our uses of them. The autonomous underwater vehicles that Bellingham and his teams have built and deployed, from the Arctic to the Antarctic, enable us to 'see' the deep reaches of the ocean in new ways that will have much to contribute to that effort."

—Franklin M. Orr, Jr., Professor Emeritus, Stanford University;
former DOE Undersecretary for Science and Energy;
board member, Monterey Bay Aquarium Research Institute

"Bellingham offers an exciting and optimistic direction for a new generation of explorers and technologists eyeing the ocean depths on Earth and among the outer planets. The agenda is challenging and long term, but technically feasible and scientifically full of promise."

—Tad McGeer, PhD; founder, Aerovel Corporation; founder, Insitu

"Comprehending the ocean, particularly the mysteries of its depths, provides the key to a healthy and prosperous planet. There's no one with a better grasp on how we can gain that knowledge—and use it ethically and wisely—than renowned marine technologies expert Jim Bellingham. His book pulls back the curtain on the hidden oceanic expanse and demonstrates why it's so vital to our future."

—George Nolfi, screenwriter, *Bourne Ultimatum* and *Oceans Twelve*; writer-director, *The Adjustment Bureau* and *The Banker*

How Are Marine Robots Shaping Our Future?

JOHNS HOPKINS
WAVELENGTHS

In classrooms, field stations, and laboratories in Baltimore and around the world, Johns Hopkins University researchers explore the world's most complex challenges and vital opportunities. The Johns Hopkins Wavelengths program brings audiences inside their stories, presenting pioneering discoveries and innovations that benefit people in their neighborhoods and across the globe in artificial intelligence, bioastronautics, cancer research, epidemiology, food systems, health equity, marine robotics, planetary science, science diplomacy, and other critical areas of study. Through these compelling narratives, their insights spark conversations in dorm rooms, dining rooms, boardrooms, and the offices of leading government representatives—including the Oval Office.

This media program—which includes narrative nonfiction books, secondary school lesson plans, digital health tool kits, and exhibits—is a partnership between the Johns Hopkins University Press and the University's Office of the Vice Provost for Research. Team members include:

Consultant Editor: Claudia Geib

Johns Hopkins Wavelengths Creative Director: Anna Marlis Burgard

Senior Acquisitions Editor: Matthew McAdam

Production Editorial Manager: Jennifer D'Urso

Senior Production Editor: Charles Dibble

Art Director: Molly Seamans

Designer: Matthew Cole

Interior Illustrations and Graphics: Nicole Kit

Production Manager: Jennifer Paulson

Publicists: Kait Howard, Core Four Media, Emi Battaglia Public Relations

Johns Hopkins University Press Executive Director: Barbara Kline Pope

Associate Vice Provost for Research: Nicholas Wigginton

How Are Marine Robots Shaping Our Future?

JAMES BELLINGHAM, PhD

with Claudia Geib

Johns Hopkins University Press
Baltimore

9 8 7 6 5 4 3 2 1

Johns Hopkins University Press
2715 North Charles Street
Baltimore, Maryland 21218
www.press.jhu.edu

Library of Congress Cataloging-in-Publication Data

Names: Bellingham, J. G. (James G.), author. | Geib, Claudia M., other.
Title: How are marine robots shaping our future? / James Bellingham, PhD,
    with Claudia Geib.
Description: Baltimore : Johns Hopkins University Press, 2025. |
    Series: Johns Hopkins Wavelengths | Includes bibliographical references
    and index.
Identifiers: LCCN 2024061666 (print) | LCCN 2024061667 (ebook) |
    ISBN 9781421450346 (paperback ; acid-free paper) | ISBN 9781421450353
    (ebook) | ISBN 9781421452920 (ebook ; open access)
Subjects: LCSH: Deep-sea sounding. | Ocean engineering—Technological
    innovations. | Robotics. | Autonomous underwater vehicles.
Classification: LCC GC75 .B45 2025 (print) | LCC GC75 (ebook) |
    DDC 551.46028/4—dc23/eng/20250421
LC record available at https://lccn.loc.gov/2024061666
LC ebook record available at https://lccn.loc.gov/2024061667

A catalog record for this book is available from the British Library.

*Special discounts are available for bulk purchases of this book. For more information, please
contact Special Sales at specialsales@jh.edu.*

EU GPSR Authorized Representative
ÐOGOS EUROPE, 9 rue Nicolas Poussin
17000, La Rochelle, France
e-mail: Contact@logoseurope.eu

On the Cover: The Next Wave of Ocean Technologies

From dugout canoes and the first sailing ships to submarines, scuba gear, and autonomous vehicles, we've developed ever more effective, efficient, and reliable means of exploring the beauty, bounty, and strategic benefits of the ocean.

Deep-sea missions are logistically complex and cost prohibitive, and those that include vehicles occupied by teams can also be dangerous; the fact that only 25 percent of the ocean's floors, mounts, and trenches have detailed mapping to guide them adds to the peril. Fortunately, the capabilities of the latest generation of marine robotics allows scientists to more deeply understand the dynamics of our interrelationship with the seas from sustenance, environmental, and mechanical perspectives—revealing science fiction worthy examples of previously unknown sea life along the way.

These cutting-edge technologies and engineering feats—building on the work of robotics pioneers including James Bellingham—allow researchers and other marine Industry professionals to push the boundaries of knowledge at lower risks and costs and in less time. One example of the new machines is on the cover: Advanced Navigation's micro autonomous underwater vehicle *Hydrus*, which is shown exploring the Ningaloo Reef, a World Heritage site in western Australia. About the size of a basketball and weighing only 15 pounds, this nimble drone can be launched and retrieved by one person and reaches areas larger vehicles can't navigate. It offers impressive data captures and efficiency to sectors across the blue economy, including aquaculture, marine salvage, energy, ocean conservation, and shipping, among others.

Contents

Preface

WE WERE LESS THAN AN HOUR into our deployment, in the startlingly blue waters off Antarctica, when we lost the robot.

We were calm, at first. But as the minutes ticked by without a signal, that sinking feeling in my stomach grew, and the ocean began to feel more and more vast and unfriendly. It was December of 1992. After nearly four years of research and development and a year of design and production, our lab team had spent the last frenzied month preparing *Odyssey*, our early proof-of-concept autonomous underwater vehicle (AUV), for this voyage—my first to Antarctica and, in fact, my very first time at sea.

Odyssey was built to be small and portable, just over 7 feet (2.2 meters) long and about 250 pounds,[1] with a shape that looked a bit like a kayak with a propeller. It had made the trip from Massachusetts down to Punta Arenas, near the southern tip of Chile, arriving just in the nick of time to board the Research Vessel (RV) *Nathaniel B. Palmer*. We then spent the two weeks of transit from South America retrofitting it, adding emergency beacons and rewriting navigation software in between standing watches and weathering fits of nausea as we

The author and the MIT AUV lab team preparing for a field program. New Hampshire's Lake Winnipesaukee provided a perfect environment to develop the skills needed to deploy and recover AUVs from the ice. A small tent over an ice hole, cut with chain saws, provided shelter for the computers and electronics used on the surface for preparing, tracking, and reviewing AUV missions. Recovery was achieved by homing the vehicle to an acoustic transponder suspended through the ice hole and capturing the vehicle in a net. The team worked in Winnipesaukee for six weeks, stopping only to ship the vehicle to Deadhorse, Alaska, for the big Arctic mission.

Photo courtesy of the author (far left).

crossed the stormy Drake Passage, preparing for the moment when it would test its mettle in Antarctic waters.

Now, just minutes into its maiden voyage, *Odyssey*'s radio beacon was silent, and it wasn't showing up on sound-based acoustic tracking. Scanning the windswept waters around our little Zodiac, the vehicle was also nowhere to be seen: In hindsight, I realized that its white-painted topside—a beacon in its home waters of Boston's muddy Charles River—might not be such a boon in the ice-scattered waters of Antarctica.

You could say that our feverish dash to the Southern Ocean was a bit of a Hail Mary opportunity. My newly founded lab, the MIT Sea Grant AUV Laboratory, had recently seen three large funding proposals denied, including a submission to the National Science Foundation for which I'd had high hopes. AUVs, which use computer programs to follow a set mission and even make independent decisions about their actions, were little more than a novelty in 1992. They had almost no history of at-sea successes, and few funders had confidence that they'd be useful. I found myself losing sleep at night, wondering how I was going to support the relatively large team of scientists and graduate students who were depending on me.

Then, along came Marcia McNutt, an MIT geophysicist, with a seemingly crazy idea: Did we want to take our vehicle, which had so far *only* operated in the Charles River, to frigid Antarctica in a month? Everyone in the lab voted against the idea—there was no way that *Odyssey* could be ready in time. But with my eyes fixed on the balance sheet, and knowing that a success could make all the difference in attracting and securing funding, I made the decision that we had to try it.

Marcia's colleagues were headed to the Antarctic to examine a "triple junction," the place where three continental plates meet, which could help geophysicists understand both the sea-floor history of the region and the broader movement of the continents. On such a trip, *Odyssey* was essentially a hitchhiker. We had explored attaching a magnetic mapping system to the AUV, but that capability was years away, so *Odyssey* was fitted with nothing more than a camera and a few rudimentary water column sensors. But the cruise was running over Christmas and New Year's Eve, a difficult time to fill bunks. Marcia offered that if I could provide warm bodies to come and stand watch, we could certainly have a few days to test the AUV.

That is, if we could keep track of it. Back on our little Zodiac, scanning the sea for our missing robot, someone offered the bright idea to call up to the bridge of the now-distant ship, which had a bird's-eye view of the scene. It was the captain who ultimately spotted the culprit: A giant petrel, an eight-pound oceanic scavenger with a six-foot wingspan, was circling and landing repeatedly on something long off in the distance. It was *Odyssey*, and every time the petrel landed on it, the seabird was pushing the vehicle just deep enough that the radio frequency (RF) beacon no longer worked—but keeping it too shallow to be picked up on acoustic pings. We zipped after the wallowing *Odyssey*, the Zodiac's motor quickly scaring off the petrel, and within the hour it was safely back on the *Palmer*'s deck.

The AUVs that journey through the world's oceans today are wildly more sophisticated than our intrepid *Odyssey*. They're equipped with satellite communications and GPS positioning, and can be fitted with instruments to collect information on nearly any ocean variable you can think of—measuring current speeds and temperatures, collecting tiny microbial life forms, or mapping huge swaths of the ocean floor. Their mission times are hundreds of times longer than *Odyssey*'s one-hour trips, and they spend these months at sea completely on their own, dropped off by a ship or launched from the shore to ply the waters independently. And while they check in periodically— sending us a "text" of sorts, like a kid off at summer camp, every time they hit the surface—we're used to these vehicles going incognito for stretches of time. The ocean is a wild place, and heavy rain, or a windy day washing waves over the top of the vehicle, can keep it from being able to talk to satellites for some time. It's not until they miss a few check-ins in a row that we start to get nervous.

But that first trial in Antarctica constituted the early days of AUV operations at sea, and probably the first AUV operations in the Antarctic. There were no rules or procedures for how to send a robot out to do science. We were, in essence, making up the rules as we went along.

Despite that first heart-stopping mishap, our trip went on from there largely without a hitch. We deployed *Odyssey* a dozen more times over our weeks off the southernmost continent and recovered her smoothly every other time. However, our first incident echoed in many nights of lost sleep, as well as in a certain disposition on my part to pace on deck and review

Odyssey under tow near Palmer Station on Anvers Island, one of three US science outposts on the continent

Photo courtesy of the author (third from left).

the lessons of each mission after *Odyssey* came home. I was unaware of the crew members gently laughing as they watched my laps on the other side of a camera above me.

Fretting aside, we were starting to get results. *Odyssey*'s camera, fitted into the bottom of the vehicle within a repurposed scuba tank, had captured images of the seafloor—coated with colorful sponges and sea stars, crawling with spiky urchins—that began to attract a lot of interest. Imagery like this was hard enough to get in warm places that were accessible to scuba divers; capturing a continuous video scan was no small task, and most researchers had never seen such images before. Scientists onboard the ship and, soon in the community at large, could suddenly imagine the sort of data that they could gain from an AUV platform.

For our lab, it was a confidence boost: the proof for our proof-of-concept, which we needed to move forward in developing this new technology—and receive funding to do so. When we had exhausted the vehicle's batteries and recovered the vehicle to deck, we joined much of the rest of the ship's complement, at Anvers Island station, which didn't abide by the dry-ship rules aboard the *Palmer*, for some celebratory drinks. I was wiped out from the preparations and stress of the operation—including writing an entirely new type of autonomous behavior for the vehicle—so I toasted our success, took a few minutes to climb part of a nearby glacier, and then returned to my bunk to crash. I was so exhausted that my memories at the end of that day were nearly psychedelic.

The cruise to Antarctica set the hook in me for life at sea that, as all sailors know, is not easily removed. During my downtime, I developed a habit of climbing the *Palmer*'s ice tower, a spire with 360-degree views for spotting potentially dangerous bergs that I found nobody ever seemed to use. I would climb up into it and read, listening to Enya's "Orinoco Flow" over and over again, surrounded on every side by dazzling turquoise and the jagged white polygons of drifting ice, plus the occasional spout of a passing whale. The glacier was calving into the bay right next to us, and the soggy ice was breaking up, leaving the surface like a giant slushy. Even when it was overcast, which, in the Southern Ocean, is most days—it was stunning; on a rare sunny day, I could not imagine a more incredible place on Earth.

Not that the weather was always so forgiving. Although the moderating effect of Antarctic ice blessed us with mostly calm weather, the *Palmer* survived its share of squalls, including a particularly rough storm just as the ship was approaching Deception Island. Sitting off the curving tail of the Antarctic Peninsula, this U-shaped volcanic island hosts an unusually protected harbor in its center. It would have been a treat to run *Odyssey* in those calm waters, both for the mission itself—as a long-standing base for both sealing and scientific research, there might even be artifacts on the bottom—and for the crew, which always relished a trip out in our Zodiac. But as mountainous Deception loomed in the distance, the weather was shifting.

The tempo of life changes with the weather when your home is at sea. As the waters swell, fewer and fewer people show up in the galley. In particularly bad weather, you might notice bandaged fingers as the impacts from unexpected waves send hands jamming into doors and skidding off metal surfaces. Little details of the ship that one might not have noticed before suddenly become very important, like the latches that keep cabinets closed or the gimbaled tables that shift with the waves, keeping food from flying away in the galley.

I woke with high hopes on the morning of the planned runs inside Deception's harbor, but once I reached the bridge, it was clear our luck was out. The wind lashed the sea surface, creating whitecap foam that streaked across translucent waves so dark blue that they were almost black. There was no question of taking our little Zodiac out in this.

Yet as I stared out, fighting disappointment, I started to notice splashes near our bow, far below the bridge. Looking closer, there were tiny, dark torpedoes, leaping cleanly in and out of the steep waves. They were penguins, swimming home or heading out to hunt, leaping like dolphins in a half-aerial and half-submersible flight through the turbulent seas. Antarctica was, as usual, offering a study in extremes: the capacity of life here to survive conditions that keep fragile humans (and their fragile equipment) confined to the safety of our homes and ships.

* * *

When most people think about oceanography, folks like Jacques Cousteau come to mind, exploring warm, clear oceans and paddling around bright coral and friendly dolphins in their swim trunks. But from the very beginning, my experience of oceanography was one of high latitudes, forbidding weather, and environmental extremes. And from that start, these extremes have inspired me to think about how technology mediates our relationship with the marine environment, even when it's at its harshest.

The Atlantic, Pacific, Indian, Arctic, and Southern Oceans, plus nearly fifty seas, all interconnect through currents to make one dynamic body of water that we collectively call "the ocean." Although it may seem to many of us land dwellers like a distant, almost alien world, the ocean is here with every one of us in our daily lives. These bonds are invisible, starting with the very air we breathe, which is largely created by tiny marine plants that float throughout the world's ocean. We owe our very existence to the ocean. This is not only because it may be the source of all life on Earth (although I think that would be enough), but also because the ocean moderates our planet's climate, keeping it comfortably warm and wet for all of that water-reliant life on board. Yet humanity's relationship with the ocean is changing.

The seas provide food and livelihoods to millions of people worldwide, and as our population grows and seeks new food sources, they have the potential to provide even more, as humans "farm" the sea in new and more extensive ways (see chapter 4). Increasingly, the seas are keeping the lights on for us at home, not only thanks to deep-sea resources like oil, but from greener sources like offshore wind (see chapter 5). And as human actions warm the planet, the ocean continues to play the role that it has since the very beginning; by absorbing both heat and carbon dioxide, its waters have staved off the worst of climate change so far—although it cannot do this indefinitely, and the repercussions of its buffering are already beginning to ripple to the surface.

Standing as the middleman in all of these links is technology, which allows humanity to operate in environments that evolution never equipped us for: from shallow surface waters to the deep ocean's toxic brine pools, bubbling methane seeps, and hydrothermal vents spewing 700-degree water. As a result, technology has enabled us to also understand the ocean's role in our planetary systems and where we fit within them. It's not just a tool for science—it also enables industries whose activities can impact marine ecosystems and contribute to greenhouse gas emissions. But after three decades in the field, I see tremendous potential for these technologies moving forward.

Those weeks in Antarctica changed the trajectory of my career. Before going to sea, I thought of my work only as a laboratory job; afterward, I understood just how important it was to experience the environment where a vehicle operated when designing it. That realization altered our lab's direction for the better. During my time at MIT, I saw scientific, military, and eventually commercial interest shift toward our vehicles, as it became clear that we were using our time at sea to make real, cumulative progress in creating a functional tool. I worked on technology that expanded the capabilities of AUVs; cultivated a team of engineers who would go on to seed an entirely new industry; and indeed developed a distant descendant of the *Odyssey* that would be spun out to a commercial autonomous underwater vehicle company called Bluefin. Later, I would move to the Monterey Bay Aquarium Research Institute

(MBARI) in California to create underwater vehicles that utterly changed the way I looked at life in the ocean, before returning to Massachusetts to found the Consortium for Marine Robotics at the Woods Hole Oceanographic Institution (WHOI). Throughout my career, I've seen marine robotics and marine technology grow smarter, hardier, and more capable, doing things we never could have imagined during those halcyon days aboard the *Nathaniel B. Palmer*.

Today, I'm the executive director of the Johns Hopkins Institute for Assured Autonomy, where I lead a number of initiatives that help my colleagues in the field successfully use autonomous technology to tackle ever-larger challenges. I've also become somewhat of an amateur historian of my field, collecting the stories of how marine exploration has evolved, especially over the latter half of the twentieth century. From this position, I believe that the ocean has the potential to provide solutions to many of the environmental and societal problems we see—and that technology can lead the way in solving them. The challenge lies in implementation: doing so responsibly, in a way that benefits not just humanity (both in the near and more distant future) but also the ecosystems we're engaging with.

There has never been a more apt time for us to have these conversations. We're approaching a future in which the ocean is ever more industrialized—when we rely on aquaculture for a greater portion of our food; when we fill our waters with wind farms and wave-energy generators to cut our reliance on fossil fuels; and when we plumb the deepest depths for minerals to power new green technology like electric cars. Many eyes are turning to the ocean to help us solve humanity's biggest challenges. But as we industrialize the ocean, we're facing the question of how we do so without entering a cycle of short-term exploitation followed by long-term environmental and social cost.

In turn, by refining and advancing marine technology, we're opening a door that can lead us to worlds beyond our own. Within our solar system, there are oceans of another kind, with salty water that swirls beneath miles of ice and seas of liquid methane bobbing with hydrocarbon icebergs. If there is life beyond Earth, this is probably where we'll find it—even if it bears little resemblance to our current understanding of what constitutes "living."

Robotics enabled us to discover that even the most forbidding corners of our oceans are rich with life. If there is life hiding beneath the ice of Jupiter's moons, or swimming around Saturn in Titan's methane seas, it's likely that we'll see it first through a robot's eyes. These robots will be even smarter and more independent than the vehicles that roam our home seas today, capable of identifying important features, examining samples, and sending that information millions of miles back to Earth. They'll be able to make decisions that help them survive in distant, hostile environments and perhaps even repair themselves when things go wrong.

That earlier me, standing on the ice tower of the *Palmer* and looking out at the Southern Ocean, might have imagined some version of this future—but at the time, it was solely within the realm of science fiction. I've always been an avid reader, and authors like Isaac Asimov and Arthur C. Clarke transported me to the alien-strewn waters on Venus, the seas of alien planets, and even to future versions of our own ocean. Although it's far from the first time science fiction has proven prescient, I've been astounded to see some of these writers' predictions come true in my lifetime. I've lived to see our oceans become populated by robots, and I imagine that my children and grandchildren may live to see the same happen to our solar system.

We stand now on what may be the very first page of that future. And while it may not seem obvious, such an interplanetary future actually has important repercussions for our own planet. The legal structures and protections that we put in place to explore new worlds, and to exploit the resources we find there, could in fact provide guidance for how we think about exploiting the nearly alien world of our own seas.

When planning out the shape and scope of this book, it was immediately clear that it had the potential to be wide-ranging. I knew that I wanted to write about the historical and scientific precipice we now stand on, in which humanity's relationship with the ocean stands to undergo a dramatic shift, as well as how we got here in the first place. Yet that book threatened to overflow to such a degree that it wouldn't fit on your bookshelves.

I've attempted to frame this relationship through the lens of technology, which has always paved the way for our land-lubbing species. In particular, this narrative focuses on how underwater vehicles and autonomous technology have opened up the oceans to science and exploration in ways we could barely have imagined a little over a century ago, and how such technology could be better deployed to the benefit of both people and the planet. It also examines how that same technology could, potentially, help us become an interplanetary species—and how that transition might impact the relationship we have with Earth.

This book is partially a work of history, partially a personal reflection, and partially a prediction of the future, all told with the help of some of the brilliant minds whom I have been lucky to call colleagues and friends. It's a call for readers to see the ocean in a new light, as a planetary partner that can help our species survive and thrive, and for a reframing of technology as a vessel that helps us to become more integrated with this environment, helping us map a responsible path to its use.

Abbreviations and Acronyms

ANGUS	Acoustically Navigated Geological Underwater Surveyor
AI	artificial intelligence
AOSN	Autonomous Ocean Sampling Network
APL	Applied Physics Laboratory, Johns Hopkins University
AUV	autonomous underwater vehicle
BBNJ	Biodiversity Beyond National Jurisdiction
BP	British Petroleum
BT	bathythermograph
EEZ	exclusive economic zone
GEBCO	General Bathymetric Chart of the Oceans
HAB	harmful algal bloom caps
HOPS	Harvard Ocean Prediction System
ISA	International Seabed Authority
JPL	Jet Propulsion Laboratory (NASA)
JUICE	Jupiter Icy Moons Explorer
MBL	Marine Biological Laboratory (Woods Hole)
MBARI	Monterey Bay Aquarium Research Institute
MH370	Malaysian Airlines Flight 370
MIT	Massachusetts Institute of Technology
MPA	marine protected area
MURI	Multidisciplinary University Research Initiative
NASA	National Aeronautics and Space Administration

NCB	Naval Consulting Board
NOAA	National Oceanic and Atmospheric Administration
NRL	Naval Research Laboratory
ONR	Office of Naval Research
OOI	Ocean Observatories Initiative
PNAS	*Proceedings of the National Academy of Sciences*
RV	Research Vessel
RF	radio frequency
ROV	remotely operated vehicle
SIMI	Sea Ice Mechanics Initiative
SOFAR Channel	Sound Fixing and Ranging Channel
SOSUS	Sound Surveillance System
SPINDLE	Subglacial Polar Ice Navigation, Descent, and Lake Exploration
SUBSEA	Systematic Underwater Biogeochemical Science and Exploration Analog
TRN	terrain-relative navigation
UN	United Nations
UNCLOS	United Nations Convention on the Law of the Sea
UNOLS	University-National Oceanographic Laboratory System
USGS	United States Geological Survey
VALKYRIE	Very Advanced Laser-Powered Kilowatt-Class Yocto-Scale Robotic Ice Explorer
WHOI	Woods Hole Oceanographic Institution

How Are Marine
Robots Shaping
Our Future?

CHAPTER 1

The Challenge of Searching the Seas

ON THE MORNING OF MARCH 8, 2014, somewhere over the Indian Ocean, a Boeing 777 airplane with 239 souls aboard vanished from the radar and satellite systems watching it. We know that Malaysian Airlines Flight 370 crashed into the ocean sometime in those morning hours, but no one can yet say why.

The mystery fueled macabre fascination and speculation over the fact that even today, in an ultra-connected world, something as massive as an airplane could simply disappear. Yet to ocean experts and to the search and rescue teams that hunted for MH370 for nearly three years, the airplane's disappearance was far from odd or conspiratorial. Rather, it was further confirmation of what we've known as long as humans have gone to sea: The ocean fiercely protects its secrets.

Just hours after the disappearance of MH370 became public, word filtered through the grapevine to technologist David Kelly. Kelly was the CEO of Bluefin Robotics, the AUV company that I'd spun out of my MIT lab, which was subsequently sold to the nonprofit science technology development company Battelle. At the time, Bluefin was in Hawaii testing *Artemis*, a 16-foot-long, 1,650-pound vehicle made to map seafloor features for oil and gas development. It was capable of diving to 14,763 feet (4,500 meters, or nearly 2.8 miles), and was equipped with multiple types of sonar that could let it spot an object like a sunken aircraft on the seafloor.

Kelly decided, over the course of just a few days, that Bluefin could pause its research and volunteered to ship *Artemis* to Australia to help with the search effort. This might seem like an extraordinary decision, especially given that *Artemis* was one of only a handful of robotic vehicles of this design in existence. However, this is surprisingly common in the underwater research realm: Our community is small and close-knit, and when disasters happen, we all feel a responsibility to lend our help.

Artemis was lowered off the side of RV *Ocean Shield* into the Indian Ocean on April 14, about 1,000 miles (over 1,600 kilometers) northwest of western Australia. It was following four pings thought to be emanating from the plane's "black box," the flight recorder that's activated in the event of a crash, which had been detected by a search aircraft earlier that month.

The vehicle spent two hours cruising downward into the utter blackness of the Indian Ocean and planned to spend the next sixteen scouring the bottom. But after completing its descent, *Artemis* unexpectedly returned to the surface. Its onboard safety system had instructed it to turn around. Although seafloor charts had shown a steady 4,500-meter depth here, as *Artemis*'s sonar followed the seafloor below, the bottom had simply fallen away from beneath it. The seafloor was much deeper than nautical charts of the area had been telling them— which, to put it mildly, was a problem (for more on the extent of seafloor mapping, see page 5).

Although we don't feel it day to day, the air in the atmosphere above us is actually exerting pressure on our bodies at all times. This pressure, as measured at sea level, constitutes "one atmosphere." In water, the power of that pressure becomes apparent. Roughly every 33 feet (10 meters) that you descend in the ocean, the ambient pressure increases by one atmosphere; this is the "squeeze" that you can feel in your ears just by diving to the deep end of a swimming pool. All of that pressure adds up fast. At the very deepest parts of the ocean, the pressure is so great that it's equivalent to the eight-ton weight of an elephant pressing on every available square inch.

Underwater vehicles are "pressure rated" for the number of these atmospheres that they can bear before all that pressure would breach their hull and cause a violent implosion. In the case of *Artemis*, it turned out that in order to reach the seafloor, the vehicle would need to descend a full 1,600 feet (500 meters) deeper than expected, to depths that engineers worried might crush it like a tin can.

Time was of the essence: An airplane's black box is designed to put out pings for 30 days, and it had been 38 by the time Bluefin got the go-ahead to deploy their vehicle. The search team had discovered an oil slick in the area on April 13 that provided another hint, but as the Indian Ocean's strong currents and the howling winds of cyclone season scattered debris further afield, the team knew that clues like these wouldn't survive the forces of nature for much longer.

The crash of Malaysian Airlines Flight 370 is perhaps the best-known modern example of how difficult it is to find anything lost to the sea; for many members of the public, it was the first time that they grappled with just how big, hostile, and poorly understood these waters can be. I'll return to the search effort for this vanished craft later; for now, keep it in the back of your mind as we move backward to consider the many leaps that humankind has taken to bridge the gap between just how inhospitable the seas can be to our species.

OPENING THE BLUE DOOR

For most of human existence, when something was lost to the sea, it was gone for good. The parents, children, and spouses of those who went to sea spent their lives in perpetual dread that their loved ones would never return.

Yet around the world, seafaring has shaped culture, trade, and empire. Starting some three thousand years ago, voyaging Polynesians set off on massive canoes, large enough to house two dozen people and their livestock, to explore and settle

Mapping the Ocean Floor

After centuries of exploration, we know that the oceans' surfaces veil many wonders. Studies of seafloor bathymetry, the underwater version of topography, reveal a world of soaring mountains and deep canyons that rival dry land's most impressive marvels. These include the nearly 7-mile (11 kilometer)–deep Mariana Trench, the planet's deepest; most of the Earth's tallest mountain, Mauna Kea, on Hawaii's Big Island—at 33,500 feet (10,210 meters), taller than Mount Everest; and the planet's longest mountain range, the meandering mid-ocean ridge, which girdles the entire planet across 40,389 miles (about 65,000 kilometers)—surfacing only briefly in the form of features that include the nation of Iceland.

Yet at the same time, as of June 2024, only about 26 percent of Earth's seafloor had been mapped in detail. While satellites can provide imaging for major features of 3 miles (5 kilometers) or more, we know little about the crucial finer details. Furthermore, maps of the ocean's floor are often of poor quality—or even flat-out wrong.

This lack of data can be dangerous, such as when imprecise charts contributed to the nuclear submarine USS *San Francisco* crashing into a seamount near Guam, crushing its front end, and injuring the majority of its crew. It hampers search efforts, as with Malaysian Airlines Flight 370 (60 percent of all flights pass over stretches of ocean with uncertain depths). Poor bathymetry stymies our knowledge of tsunami paths, can compromise energy and military needs, and limits our ability to protect at-risk marine ecosystems and habitats—and learn about as-yet unknown species.

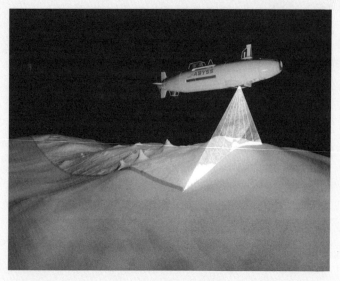

Image courtesy of T. Kwasnitschka and N. Augustin, GEOMAR Helmholtz Centre for
Ocean Research, Kiel

Oceans cover about 70 percent of the planet's surface, so scale
alone works against mapping efforts. Added to that challenge are the
intense pressure, visibility issues, and extreme cold that can bedevil
scanning equipment.

Seabed 2030—a mapping initiative led by GEBCO, the Nippon Foundation, the International Hydrographic Organization, and UNESCO—
aims to complete mapping of the entire seafloor, to a level of detail
of at least 320 feet (100 meters), over the next five years. To do so,
they're calling upon a number of techniques and technologies, and
are also seeking to access existing data from government and indus-

try organizations that have information archived in their servers—at least, that which can be revealed without risks to military and other top-secret interests.

Global powers are also turning their attention to the seabed. China has been facing fire for sending seafloor-mapping ships well outside their national waters, such as into the Indian Ocean. Beijing claims its interests in the seafloor are for science and potential deep-sea minerals, but this data could also help Chinese submarines evade detection.

Meanwhile, electrical engineers, material science experts, audio engineers, and imaging specialists, among many other inventors and technicians, continue to increase the capacities of deep-sea imaging and exploration.

FURTHER READING

Mizokami, Kyle. "15 Years Ago, a U.S. Navy Submarine Ran into a Mountain." *Popular Mechanics*, January 8, 2020. https://www.popularmechanics.com/military /navy-ships/a24158/uss-san-francisco-mountain-incident/.

National Oceanic and Atmospheric Administration, Ocean Exploration. "What Is a Mid-Ocean Ridge?" https://oceanexplorer.noaa.gov/facts/mid-ocean-ridge.html.

The Nippon Foundation / GEBCO, Seabed 2030. "Accelerating Ocean Mapping." https://seabed2030.org

across an area of more than 100,000 square miles (nearly 26 billion hectares), the vast majority of which was ocean. These explorers found their way to tiny specks of islands using only stars and clouds, the ocean's swell, the color of the sky, and the movement of sea life.

On the other side of the world, Norwegian Vikings in longboats made landfall on the shores of eastern Canada before Christopher Columbus was a twinkle in his great-great-great-grandparents' eyes, and started a cod fishery off its shores that lasted into the 1700s.[1] Ancient Greek sailors and Sumerians, leaving from what is now Iraq, crisscrossed the Indian Ocean in square-rigged sailboats and oar-equipped triremes to trade with India. Chinese empires conducted a bustling trade with the rest of Asia from the 600s onward, reaching all the way down to southern Indonesia, and in the eleventh century, introduced the first mariner's compass to the world.[2]

Additionally, there have always been those who broke the surface to search the depths below. Many cultures have a long tradition of diving to collect food and treasure. Notably, the Japanese *ama* ("sea women") have a centuries-old culture of free diving for pearls and seafood; a few of these women are scrambling to preserve this practice in the twenty-first century. In Southeast Asia, the Bajau people have actually developed larger spleens that help them hold their breath longer while hunting marine life.[3] Human vision is extremely limited beneath the surface; water is around the same density as the fluid in our eyeballs, so while underwater, light doesn't refract into the eye in the same way our brains evolved to interpret it through air. Yet among these cultures and others who dared plunge below the waves, our eyes were the first tools used to search the sea.

Over time, scientists and mariners developed tools that could measure other parameters of the sea: tools like sounding weights, weighted and knotted lines that roughly measured depth; the Arab-invented *kamal*, a line attached to a wooden card that provided a rough estimation of latitude; or the mariner's astrolabe, which calculated angles to the sun or a known star to determine latitude. Even so, for a long time the oceans remained difficult places for us to explore. There's the simple fact of the environment itself—wind and waves are a constant, and the weather can change quickly and violently. Below the surface, currents bustle along even on the calmest of days, sometimes at speeds that can quickly sweep something away. The deeper you go, the more complicated it becomes for human bodies and machinery alike, as pressure mercilessly increases. Yet even working in shallow water is hard: Salt wears away our toughest materials; opportunistic plants and animals latch onto any surface left below the waterline too long; and, of course, water itself has a talent for finding its way into anything and everything, no matter how carefully sealed.

Even if you can overcome these issues, water's intrinsic qualities also make it difficult for any instrumentation that we develop to extend our land-evolved sight. Light behaves differently when it moves through water. It travels more slowly than in air and refracts—changes direction—when it passes through the ocean's surface. It is scattered by water molecules and floating particles such as sediment. As a result, sunlight only makes it about 320 hundred feet (100 meters) down into the ocean before it surrenders to darkness, or less in murky coastal waters.

Even color doesn't behave the same in water. Water absorbs and scatters certain wavelengths of light: Short wavelengths, which create colors like blue, green, and violet, can penetrate as deep as light goes, but long ones like yellow and red cannot. A strawberry, floating 50 feet (15 meters) down, would appear a washed-out blue-gray in ambient light.

Sound, too, behaves oddly below the ocean's surface. Sound travels faster and farther in water than in air; that's why, if you're snorkeling or scuba diving, you can hear the drone of distant boat engines even when they're nowhere near you. The more dense water becomes, the faster sound will move. As a result, sound travels fastest as temperature, salinity, and especially pressure increase, all of which make seawater denser. That also means that sound changes direction when passing through areas where any of these parameters change, making the ocean the acoustic equivalent of a house of mirrors.

Some early scientists made inroads to understanding this environment. Many cultures developed a sufficiently sophisticated understanding of winds and currents to be called early oceanographers. During the fourth century BCE, Aristotle notably studied and described more than a hundred fish species as well as cephalopods, sharks, and skates. He has been called both the world's first marine biologist and its first biologist in the modern sense.[4] Yet for a long time, the widespread view of the ocean, as a habitat and an environment, was that it was the home of sea monsters and demons, supernatural occurrences, and vengeful gods.

From the 1400s onward, Europeans entered an era of exploration and colonization, and once again widely held views of the ocean changed. There are too many voyages and too many important discoveries[5] in this period to even approach naming them all, but the starting edge of modern Western marine science is often attributed to the three voyages of Captain James Cook, between 1768 and 1779.

Cook was among the first explorers to bring a dedicated naturalist on each of his journeys, and these men collected numerous samples of marine plants and animals along the way that were previously undocumented by science. Cook himself was a talented navigator and mapper, and he filled in great portions of the world's surface map with reasonable accuracy.[6] His successes can be attributed in part to a new technology: the marine chronometer, created by a clockmaker called John Harrison, which used a spring to keep accurate time despite the constant motion of a ship. A ship's latitude could easily be determined using the position of the sun; but with the use of the chronometer, Cook and other navigators could finally determine their longitude, allowing a fairly accurate fix on their position. This allowed for shorter, more efficient voyages—not getting lost really speeds things up—and safer navigation around known hazards.

These voyages shaped the public's perception of the ocean as something that could be understood: mapped documented, and neatly described in books and lectures. This is the era in which Charles Darwin set off for the Galápagos Islands aboard the *Beagle*, the voyage that would inspire his world-changing theory of evolution—thanks, in part, to his observation of marine species like barnacles.[7] Still, this period of "science-ing" the sea led to some serious misunderstandings. In the mid-1800s, British naturalist Edward Forbes concluded from a survey of the Aegean Sea that there was no life in any of the oceans deeper than 300 fathoms, a measure equal to around 1,800 feet. This theory came to be known as the azoic (from Latin, meaning "without animals") theory, and it caught on widely.

Once again, technology played a role in this development, in this instance for the worse. As author Helen Scales points out in her book *The Brilliant Abyss,* Forbes's main tool was a canvas bag dotted with small holes, which quickly filled with mud when dragged across the seabed. Few organisms will show up when your net is completely clogged. Additionally, the Aegean Sea was perhaps the worst model to use as a standard for the global ocean; it's actually known for being relatively devoid of life, thanks to exceptionally nutrient-poor surface waters.[8] Forbes's theory was like making conclusions about a rainforest based on samples from a desert.

Evidence already existed that the azoic theory was false. But to naturalists at the time, Forbes's idea seemed logical. It neatly mirrored the habitat zones that scientists were discovering at different altitudes, with life often growing scarcer towards high mountain peaks. They knew of the immense pressure in the ocean's depths; nobody had yet been able to develop even a simple device, such as a thermometer, that could survive the trip. "The depths of the ocean are quite as impassable for marine species as high mountains are for terrestrial animals," naturalist Louis Agassiz wrote in 1851. "Not only are no materials found there for sustenance, but it is doubtful if animals could sustain the pressure of so great a column of water."[9] (This would hardly be the last time he got the science wrong: Agassiz also opposed evolutionary theory and was racist, defending slavery by ranking different races of humans as different—and inherently unequal—species.)

It took until the late 1800s for the azoic theory to fall out of favor. This was thanks first to the work of British Royal Society expeditions aboard the ships HMS *Lightning* and HMS *Porcupine*, which hauled up sea stars, sponges, and corals from as deep as 650 fathoms (nearly three-quarters of a mile, or around 1.2 kilometers).[10] Then, between 1872 and 1876, a small British warship called the HMS *Challenger* circled the globe on the first dedicated oceanographic expedition—in part to build on the discoveries of the *Lightning* and *Porcupine* and settle the azoic theory once and for all. The *Challenger* expedition, as it would become known, spent those four years sampling the ocean floor, collecting marine animals, and mapping the depths and features of the bottom. Their work not only put the final nail in the coffin of the azoic theory—they collected their deepest organisms from below 2,000 fathoms,[11] more than two and a quarter miles (3.7 kilometers)—but it heralded a new era in marine science. The *Challenger* naturalists systematically collected depth, density, and temperature measurements wherever they went, and mapped currents all around the globe. In doing so, the *Challenger* expedition painted the first picture of the ocean's structure and inner workings.

Such discoveries would form the basis of physical oceanography,[12] a field that would come to play a role in many of humanity's most inspiring advancements. But first, it would need to grow, and it would do so through some of our species' most violent acts.

CREATION THROUGH DESTRUCTION

The concept of a vessel that can operate completely underwater has long captured the minds of those who create and

Layers of the Ocean

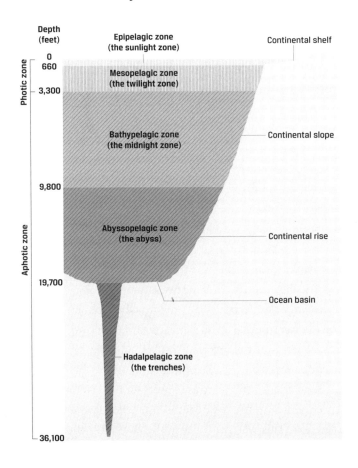

Depth (feet)

Photic zone

Aphotic zone

Epipelagic zone (the sunlight zone)

Continental shelf

0
660

Mesopelagic zone (the twilight zone)

3,300

Bathypelagic zone (the midnight zone)

Continental slope

9,800

Abyssopelagic zone (the abyss)

Continental rise

19,700

Ocean basin

Hadalpelagic zone (the trenches)

36,100

of those who wage war. Leonardo da Vinci sketched out ideas for a crewed underwater weapon back in 1515 but drew the device in a disassembled form; some historians believe he made these plans intentionally vague, owing to concerns about the potential dangers of such an idea falling into enemy hands.[13] It took a few centuries, but da Vinci's concerns eventually became reality. In 1863, during the American Civil War, Confederate soldiers in the HL *Hunley*, a primitive submarine fashioned from an iron steam boiler, successfully sank the warship USS *Housatonic* by attaching a torpedo to her side. The *Hunley* sank as well, killing everyone on board and living up to its nickname of the "peripatetic coffin." But this Pyrrhic victory was the beginning of a new kind of warfare.

In the year 1900, the United States Navy commissioned its first working submarine, the USS *Holland*;[14] in that same year, the British Royal Navy ordered five submarines from the American company Electric Boat. In 1903, the German Imperial Navy began stocking up on submarines of their own. As a consequence, by the time World War I began in 1914, several of its major players were capable of waging war underwater.

Submarines, referred to as U-boats for their German name, *Unterseeboots*, were used to devastating effect as World War I raged on. This was particularly the case after the German Imperial Navy began targeting civilian ships, such as the infamous sinking of the *Lusitania*—one of the major contributing factors in the United States' decision to enter the war. However, this tragic era also steered the direction of scientific research.

Prior to World War I, it was difficult to make a living as an American oceanographer. Oceanography had flourished in Europe in the early twentieth century, where many countries recognized its practical application to the health of fisheries, central to several countries' economies. In 1902, Denmark, Germany, Britain, Sweden, Holland, Finland, Norway, and Russia founded the International Council for the Exploration of the Sea (ICES), largely due to concerns over the dwindling of fish stocks in the North Sea. Author Gary Weir writes that the "practical applications and economic value attracted talented people, stimulated new perspectives, and prompted generous public sponsorship" for European oceanographers.[15]

This was not the case for ocean science across the pond. Just two American oceanographic research institutions existed at the time: the Scripps Institution of Oceanography in Southern California, and the Marine Biological Laboratory (MBL), in Woods Hole, Massachusetts. Before the war, they largely focused their work on the seashore and coastal waters, as they owned no ships and so had no way to access the deep sea. Both had few employees and minimal funding; MBL did not even keep staff on year-round.

All of that changed dramatically during World War I. As early as 1914, the Navy recruited Allied scientists into defensive anti-submarine research. This research produced such innovations as the magnetic submarine detector, invented during this period by a young Vannevar Bush—a physicist who would go on to play a role in the Manhattan Project, the creation of the National Science Foundation, the founding of Raytheon, and the early research that led to computers. (Bush's device worked, but only if used aboard a wooden ship.)[16] When the United States joined the war in 1917, the Naval Consulting Board (NCB) created special committees for the scientific study of wartime problems, including the possibility of improving the detection of unseen submarines.

Prior to the war, in the great era of steam-driven ocean travel, scientists had already realized that ships needed a better way to monitor their environment. The sinking of the *Titanic* in 1911, its hull split by an iceberg that loomed out of the night, signaled this need, and the attacks of German U-boats made it brutally apparent. Threatened by both nature and enemies, researchers discovered that they could detect objects in the water using pings of sound. By calculating the time it took for that sound to bounce back to the source, one could figure out the distance to the object it bounced off of, a use of sonar known as echo ranging.[17] The principles are the same as the echolocation that bats and whales use to find their prey; to any student of the marine world, there is some irony in the fact that it took tremendous effort for humanity to develop something that a whale can do from the moment it's born.

In 1917, the UCB recruited the Submarine Signal Company in Massachusetts, which by that time had developed a functional iceberg detector using sonar and demonstrated that they could measure the ocean's depths with these acoustic pings. They couldn't quite nail down a working U-boat detector, though. So the NCB created a cooperation between the Submarine Signal Company, General Electric, and Western Electric, a collaboration that would prove fruitful: It produced effective echo ranging devices and working hydrophones, underwater listening devices that could be used to detect approaching subs.

By the time World War I ended, in 1918, American leadership not only saw the power of undersea warfare—they also saw just how powerful ocean science could be in shaping it. The United States manufactured three thousand of the Submarine Signal Company's hydrophones for use during the war, which proved highly effective in detecting U-boats as they approached Allied convoys, preventing an unknowable number of sinkings.[18] (Outside of shooting at submarines at the surface, vessels had little ability to fight back until depth charges, underwater bombs that could damage and even sink U-boats, were implemented during the latter half of the war, between 1916 and 1918.) Even so, by the end of the war German U-boats alone had sunk more than five hundred ships, an estimated two hundred of them American,[19] at the cost of more than fifteen thousand lives.[20]

While the conflict had ended, in the decade that followed, American military minds anticipated that the "Great War" wouldn't be the last—and that during the next global conflict, even more of the fighting would happen below the waves. In the lead-up to World War II, they began funneling unprece- dented amounts of money into oceanographic research work. This, in turn, sparked the growth of new labs, new academic programs in oceanography and engineering, and wider interest in this burgeoning field.

Navy funders were liberal in scope, often supporting basic research that seemed only tangentially related to warfare. For many years, this meant that the Navy avoided any stain of bias and intellectual restriction. Yet that didn't mean they were free of either. In her excellent history of the military's funding of oceanography, *Science on a Mission,* Harvard University historian Naomi Oreskes lays out a convincing case that this no-strings-attached view of Navy funding is not quite true. The Navy's interest influenced which projects oceanographers chose to pitch and what areas they chose to work in; biology, for example, was largely ignored, as the Navy prioritized fields, like physical oceanography, more directly tied to submarine warfare. Military secrecy also meant that some topics were simply impossible to investigate. Notably, Oreskes argues that the theory of plate tectonics was likely delayed by nearly thirty years, as the Navy labeled almost all seafloor bathymetry data as classified.

Nonetheless, Navy-supported research significantly improved our understanding of how the ocean worked, from the global movement of tides[21] to the basic shape and depths of the seafloor. Among the most important outcomes of this work was a greater understanding of how sound works in the ocean. In June 1922, Navy physicist Harvey Hayes used a sonic depth finder—developed out of his wartime research into active sonar—to make the first complete bottom profile of the Atlantic during a crossing from Rhode Island to Gibraltar, revealing the shapes and depths of the ocean in a stripe along his ship's path with an unprecedented level of detail. His results, Weir writes in the book *An Ocean in Common,* "went through the scientific community like a bolt of lightning"— and both civilian scientists and naval minds alike saw the incredible potential of such a device.[22]

In the 1930s, the newly founded Woods Hole Oceanographic Institution (WHOI)—at the time, a small, poorly funded lab in Massachusetts that only employed staff during the summer— began investigating what sonar operators called the "afternoon effect." This term referred to the strange fact that sonar equipment worked well in the morning but poorly in the afternoon, particularly when it was warm and sunny. The researchers discovered that rising temperatures at the ocean's surface were slowing the velocity of sound in water, and that sunny afternoons were not the only time that this effect confounded sonar operations.[23]

Temperatures rapidly decline in the meters below the ocean's surface, a region of change known as the thermocline. As a result, sound moving through this area can be refracted down so dramatically that it misses a shallow target, like a submarine patrolling just below the surface. In order to ad-just, researchers turned to a collaboration between WHOI and another newly founded organization, the Naval Research Laboratory (NRL), to conduct new research on depth and temperature that allowed sonar operators to adjust.[24]

It wasn't long before war came again, and this research be-gan to have real-world, life-or-death implications. The Navy had been correct in expecting that World War II would be a war fought underwater. In its early days, this war appeared to be one the Allies were not winning. On the western front, the destruc-tion wrought by German U-boats on transatlantic supply routes earned the nickname "the Battle of the Atlantic." Simul-taneously, German U-boats terrorized the Gulf of Mexico,[25] seeking to disrupt Allied access to oil and succeeding; the Ger-mans sank fifty-six Allied ships in the Gulf, twenty-eight of which were oil tankers. In the Pacific, too, the Allies had a rocky start: Between 1941 and 1943, defective torpedoes greatly limited the success of US submarines.[26] (Although Germany didn't invent the torpedo—an American and a Brit share that accolade—they had carefully refined the technology after World War I, working in secret to avoid the restrictions im-posed by the Versailles Treaty.)

Yet by the end of the war, American submarines had effec-tively disabled the Japanese Navy. American submarines alone would be credited with sinking about half of the Japanese ton-nage that ended up on the bottom of the Pacific, as well as with rescuing hundreds of downed aviators from the ocean's clutches.[27]

What changed? One of the deciding factors was increased knowledge of the undersea environment. The case of the torpedoes was one such factor: With careful testing, it turned out that one reason these missiles were exploding too quickly was because their magnetic sensor, meant to detect a ship's steel hull, was responding to the largely unstudied variations in the Earth's magnetic field. By switching to contact-triggered exploders instead, American torpedoes went from dud to deadly.

The study of sound also brought an enormous payoff. Researchers had learned early on that radio waves attenuate quickly underwater. This made it essentially impossible for a ship to use standard radio channels to communicate with a submarine, or for two subs to communicate with each other, unless they were on the surface. Research on the properties of sound in the ocean allowed the Navy to develop the underwater telephone—also known as Gertrude, a nickname whose origin has been lost to history. Gertrude messages convert a voice message or Morse code into a high-pitched sonar signal, allowing it to travel through water.

But one of my favorite illustrations of how ocean science changed the course of war hangs on the wall of the port office on WHOI's seaside campus. Scrawled in untidy cursive, it's a note briefly mentioning Allyn Vine, one of the oceanographers who took up research on the thermocline for the Navy. Vine ultimately designed a version of the bathythermograph (BT), an instrument capable of measuring water temperature at various depths, that could be used on submarines, allowing them to locate the thermocline.

The note on WHOI's wall comes from the log of the USS *Guitarro*, a U-boat operating in the Pacific that was later decorated with multiple battle stars for its service. It describes an incident in which the *Guitarro* was damaged by anti-submarine countermeasures and found itself chased and harassed by Japanese ships. As the ships closed in, the officer was able to steer the sub under a sharp thermocline layer, where they found the attacks became "less and less fruitful" as they vanished off the ship's sonar. The officer then writes, "My sincere thanks to Allyn Vine of Woods Hole Inst. for the time he spent explaining the value of BT observations to me." Understanding how the ocean worked had directly saved 108 men's lives.

Little surprise, then, that the Navy exited World War II committed to continuing its support for oceanographic research. And while outright warfare had ended, military interests in the sea had not. As the uneasy postwar peace between the United States and the Soviet Union shifted into the simmering hostility of the Cold War, monitoring Soviet submarines now prowling the Atlantic became critically important to Western powers. To support this effort, the US Navy started becoming interested in new types of underwater vehicles beyond submarines—vehicles that could gather information and perform complex operations in the deep sea. This interest would fuel the development of an entirely new industry in ocean science, one that would extend marine technology beyond its military roots, opening new frontiers in science and discovery.

CHAPTER 2

Opening the Door

SEPTEMBER 9, 1959. It was two years after the Soviet Union had launched a disco ball–sized metal sphere called Sputnik into orbit, propelling the United States into a nationwide effort to be the first to send humans off-world. The Space Race was on with America's greatest rival, but on September 9, the US contender was missing in action.

The missing space capsule was known as Big Joe, an unoccupied, lightbulb-shaped vessel roughly the size of a minivan. Project Mercury, NASA's first human spaceflight program (founded in 1958), had launched Joe in an arc through the atmosphere that morning from Cape Canaveral, Florida. The aim was to test the heat shields that they hoped would one day protect astronauts from the enormous heat of reentry, plus the capsule computer's ability to control that reentry and the recovery team's ability to find it after splashdown. But shortly after launch, signals from the Mercury capsule had gone haywire and then vanished entirely.

Several excruciating minutes later, a report came in from the Mission Planning and Analysis Division. In an oral history interview, Robert F. Thompson, head of recovery operations for NASA's Space Task Force Group, remembered the report going something like this: "It left here, but we don't know where it went."[1]

Later, the team would discover that the Atlas rockets propelling Big Joe skyward had stayed attached to the capsule longer than they were supposed to. As a result, the weighed-down capsule was going 1,800 miles (2,897 kilometers) per hour slower than planned when it reached the edge of space, 87 miles (140 kilometers) above Earth, and prepared to turn around and start falling again. When Big Joe finally started to come down—completely out of fuel after its primitive onboard computer had tried, in vain, to correct the error—the capsule was screaming toward Earth 500 miles (804 kilometers) closer to Florida than calculations said it should be.[2]

But some people knew where the capsule landed, because they could hear it.

In fact, three someones were listening for Big Joe: Three waiting Navy ships had sonar operators with their ears to the water, and around the time Joe hit the Atlantic, they heard a very distinctive set of sounds from beneath the sea—a series of fuzzy pulses, growing sharper and closer together until they culminated in a sharp click.[3] As the Mercury capsule deployed its parachutes, it had dropped a fist-sized grenade into the ocean, which sank to a depth of 3,000 feet (914 meters) before automatically exploding, creating the sound the sonar operators heard. The depth where this device detonated was key. At that depth, any sonar operators listening to the ocean within nearly a thousand miles would have been able to hear the explosion, and by coordinating with each other, get an accurate fix on the location of the capsule. Within a few hours, the Big Joe capsule was floating on a Navy vessel alongside the docks in Puerto Rico, ready to head back to Florida to propel the Space Race onward.

Big Joe's savior was a feature of the ocean dubbed the Sound Fixing and Ranging (SOFAR) Channel. In the 1930s, oceanographers working for the Navy had discovered that sound's speed underwater increased at higher pressures, but decreased as temperatures fell. In the 1940s they found that there is a sweet spot in the oceans—varying by location, but usually between 2,400 and 3,000 feet (731 to 914 meters) in temperate latitudes—where temperature and depth are perfectly balanced to let sound travel at the slowest possible speed in seawater. As a result, sounds in this channel can propagate thousands of miles, crossing entire ocean basins under the right conditions.

Naomi Oreskes calls this discovery "the most important wartime development in oceanographic science." Indeed, oceanographers immediately recognized its importance to the war effort. By listening to the SOFAR channel, one could pick up the propeller of a submarine from hundreds of miles away, and with multiple listeners, triangulate its position. Maurice Ewing, one of the WHOI scientists who made the discovery, proposed and tested another use of the SOFAR channel: developing small devices that would explode, either through implosion or a timed fuse at the depth of the SOFAR channel, allowed downed pilots to signal their location to listening stations nearby and be rescued.

Ewing perfected the device, called a SOFAR bomb, only months before World War II ended, so it was never deployed during the actual conflict.[4] But in the years that followed, the scientific community found plenty of use for these discoveries. SOFAR bombs were deployed on all of the Mercury Program's uncrewed spacecraft and two of the flights with humans aboard. The SOFAR channel itself was initially used to monitor submarine movement during the Cold War, but later became an important tool for studying marine mammals, deep-sea earthquakes, and the effect of man-made noise on ocean life.

This is the trend that marine science followed in the second half of the 1900s. Many discoveries and innovations made by civilian scientists working under military funding enabled extraordinary leaps forward in our understanding of the ocean: in oceanography, marine biology, undersea technology, and in space travel, which intersects with ocean science more than most people may realize (see chapters 7 and 8). With each leap, scientists saw that a better understanding of the ocean—and better technology to study and explore it—could be useful across the board.

THE SEARCH FOR *THRESHER*

Amid escalating tensions during the Cold War, the Navy was already interested in expanding their access to undersea vehicles. At the time, the military was captive to systems like the then-new nuclear submarines, which are expensive to both build and operate and are not well suited to activities like search and recovery, especially at great depth. Could smaller, more agile systems offer a competitive edge without such high operational costs and perhaps with fewer risks to human lives?

The Navy began investing funds in underwater vehicles at WHOI and Scripps shortly after the war ended. Their initial hope was to develop machines that could be used to service the new network of deep-sea hydrophones installed in the SOFAR layer in the 1950s, known as SOSUS (Sound Surveillance System). But it was an accident that highlighted the true need for such a vehicle, and that set the stage for civilian science to gain access to a new realm of ocean research.

In April 1963, a military submarine called the USS *Thresher* sank in the Atlantic, east of Cape Cod, while performing tests at a depth of around 730 meters (2,395 feet, or just under half a mile). One of the last communications that nearby Navy vessels heard before the *Thresher* vanished off comms was a garbled message sent by underwater telephone: "Minor difficulties, have positive up-angle, attempting to blow." One lieutenant later reported that five minutes later, he heard a sound familiar from his service during World War II: the dull thud of a ship's interior spaces breaking apart. Then there was silence.[5]

As the hours, and then days, passed following the *Thresher's* disappearance, the Navy prepared to change their operation from search and rescue to marine salvage; with the grim discovery of floating debris from the sub, the search turned definitively to the seafloor.

The *Thresher* had sunk in an area where the seafloor lay at a depth of 2,300 meters (8,400 feet), more than a mile and a half down. Ships were able to find some scattered remains of the sub—and, importantly, verify that its nuclear engine was not leaking—using towed sleds that carried magnetometers and Geiger counters, as well as cameras that could take fleeting photos of the bottom. But in order to figure out what had happened to sink the *Thresher*, the Navy needed more than the few frames gathered from a drive-by.

The Navy attempted to locate and study the submarine's wreckage using a vessel called the *Trieste* and its successor, *Trieste II*. These two vessels were bathyscaphes, predecessors to deep-diving submersibles. Designed by Swiss oceanographer Auguste Piccard, the *Trieste*'s shape was similar to that of a submarine. However, instead of humans living and moving about in the great horizontal cylinder at its top, they were crammed into a tiny pressure sphere perched at the bottom. The upper cylinder was in fact an enormous float filled with gasoline, which made the bathyscaphe buoyant. Iron pellets, held in place by a

The *Trieste*, circa 1958–1959.

Source: US Naval History and Heritage Command

magnet, made it heavy enough to sink to the ocean bottom. When the *Trieste* was ready to return to the surface, the magnets would drop that iron ballast and the bathyscaphe would rise.

At the time of the *Thresher*'s disappearance, this was the only vessel capable of descending to the depths where the sub was lost; already, *Trieste* had accomplished the unthinkable by descending to Challenger Deep, the deepest point in the ocean, 6.8 miles (10.9 kilometers) down within the Mariana Trench. In the summer of 1963, following photographs of seafloor debris and oceanographers' models of where the sub had likely drifted after losing control, the *Trieste* began its search.

But the *Trieste* was made for the simple goal of reaching the depths, not working in it. The bathyscaphe had never been equipped for a deep-sea search, even after a 1964 refit, which streamlined the submersible, added a mechanical arm, and gave it the title *Trieste II*. Its complex ballasting system made the bathyscaphe difficult to position with any level of precision, and it could barely move fast enough to fight the ocean current at the bottom; its tiny propellor puttered along at around three-quarters of a mile per hour. Once the *Thresher* was located, two of *Trieste II's* dives to investigate it, each expensive and complicated to recover from, failed completely; on the third, the mounted camera didn't work.

Ultimately, the *Trieste II* did land on the shattered remnants of the *Thresher*, quite literally; during the fourth dive, the pilots discovered that they were sitting not on the seafloor, but on the sub's broken, partially mud-covered hull. Thanks to detailed pictures of these remains and studies of other submarines like the *Thresher*, the Navy eventually concluded that a leaking pipe had likely caused rapid flooding, causing the sub to begin sinking and shutting down its reactor, which cut off propulsion. As the sub sank below the depth it was built to withstand (called its "crush depth"), it imploded.

But it took almost a year in total to find and photograph the sunken sub, and the Navy walked away feeling rudely awakened to its inadequacy in deep-sea operations. Frank A. Andrews, a Navy captain and submarine group commander who led the search, wrote in 1965 that the search "demonstrated only too clearly the degree of ignorance and inability which surrounded the entire business."

ENTER *ALVIN*

By the time of the *Thresher* search, the Navy had already attempted to fund the construction of one crewed undersea vehicle, the *Aluminaut*. The plan was that the Office of Naval Research (ONR) would fund the construction of the vessel and rent it to WHOI for three years, which would fulfill the office's mandate of funding scientific research. After that period, ONR would take ownership of the vessel and use it to service their new SOSUS array off Bermuda.

But the plan seemed to become entangled at every step. Allyn Vine, the WHOI oceanographer who had been so instrumental to Navy sonar efforts during World War II—and who now led the effort to get WHOI access to a submersible for scientific use—strongly disagreed with the first design from the Reynolds Metal Corporation, which was building the sub, as their blueprints included no windows in the piloting station. (Yes, that's the same Reynolds company that makes kitchen tinfoil; they were seeking, at the time, to promote the utility of aluminum.) Even when more windows were added, the WHOI and ONR teams remained concerned about the safety of the vessel: Aluminum had not yet been tested against the pressures of the deep sea, and tests kept coming back with cracks and ruptures in the prototype cylinder. This didn't help with resistance to the project within WHOI; behind the scenes, staffers were starting to express skepticism that the vessel would even be useful for their scientific work.

Additionally, Reynolds was asking exorbitant leasing fees for WHOI to use the vehicle, which seemed to increase with every passing month. And in the cherry on top, the chairman of that corporation, Louis Reynolds, didn't actually want to transfer ownership of the vessel back to the Navy at the end of its construction. He wanted to keep it, reap the benefits of owning what he saw as a historic vehicle, and donate it to the Smithsonian at the end of its life.

By 1963, WHOI and the ONR had had enough with Reynolds; they decided to put out a new bid for the construction of a different research submersible, one with more stringent safety and materials requirements on top of "good control and maneuverability . . . especially for bottom work." General Mills took the contract—in another kitchen crossover, the cereal company briefly had an Aeronautical Research Division and Electronics Division. Then, the *Thresher* sank, and the Navy was forced to confront the fact that their previous plans for a submersible program at WHOI might be more important than they had realized.

A new Deep Submergence Systems Review Group, headed by the Navy but including both ONR engineers and WHOI scientists, recommended that the Navy develop a long-term program for WHOI to rent the new General Mills vessel, rather than use it for short-term contracts before returning it to the military. Doing so ensured that the vessel would remain in working condition, as machines that sit idle rarely work when you need them. It provided the added bonus of building a base of expert users that might come in handy to the Navy one day, as well as more diverse funding opportunities.

The Navy also recognized that while this longer-term relationship meant that it would have the submersible available any time an emergency like the *Thresher* arose, such emergencies should be few and far between. As a result, the rear admiral heading the group recommended that the vessel "should be conceived against its probable background usefulness in other phases of science and engineering." Rather than a Navy vessel that occasionally was used for science, WHOI was getting a science vessel that the Navy might occasionally borrow.[6]

In 1964, Woods Hole got *Alvin*—named for Allyn Vine, whose tenacity in pushing WHOI toward developing a submersible became one of his legacies within the field. *Alvin* gave the Navy everything it had been missing from the *Trieste*, and gave scientists an incredible platform for deep-sea research: It was faster and more maneuverable, could dive to 6,000 feet (1,829 meters) and was capable of performing a variety of different operations on the bottom using a mechanical arm. The submersible proved its usefulness almost immediately after it was commissioned: In 1966, it located one of four hydrogen bombs that dropped into the sea off Palomares, Spain, after an aircraft collision.

The Navy kept the vessel's schedule busy with service calls to the SOSUS array off Bermuda. And despite WHOI's initial skepticism, proposals from scientists began to pile up, as the community eagerly anticipated the role it could play in research. It would take time for many of these projects to materialize; *Alvin* was, and remains—it's still diving all these decades later, although after several refits—expensive to operate. Over the next decade, it took the help of government funding agencies to make it possible for the scientific community to afford its use.

Even so, more submersibles would follow *Alvin* from around the world, popping up in France, the Soviet Union, and Japan. At the start of the 1960s, only four existing human-occupied vehicles (HOVs) could operate at depths of 650 feet (200 meters) or more; by the close of the year 1970, there were more than a hundred worldwide.[7] At Scripps and within Navy laboratories, engineers were working on remotely operated underwater vehicles; meanwhile, throughout the 1960s, the Navy sank three undersea habitats off the coast of Bermuda and Southern California called Sealabs, which proved that "aquanauts" could live and work for extended periods at hundreds of feet below the ocean. It seemed like only a matter of time before humanity had some permanent occupancy beneath the waves. The buzzword within the community was the potential creation of a "Wet NASA," a marine counterpart to our growing space presence.

Marine scientists began pressuring Congress and the federal government to invest accordingly in civilian ocean activities. In October 1970, the United States created the National Oceanic and Atmospheric Administration (NOAA), which brought the pieces of ocean science that had been scattered across nearly two dozen agencies under one roof.[8] However, by the time NOAA came into existence, in the early days of Richard Nixon's presidency, the tides had already shifted. The US economy was in decline, and the agency was far from the new administration's main priority. Although the concept of an aquatic NASA would continue to resurface over the decades that followed, there was never enough funding nor momentum to make it a reality.

THINKING FOR THEMSELVES: ROBOTS' EARLY DAYS

Even as the federal government's appetite for funding big ocean projects diminished, smaller options moved in to fill the gap as technology progressed. In the late 1960s, Scripps researchers developed Deep Tow, a towed vehicle with a package of sonar, cameras, and geophysical sensors aboard. This primitive re-

Living on the Seafloor

During the 1950s and '60s, as humankind began exploring space, researchers and adventurers also turned their sights on the inverse: living and working underwater. Although scuba allowed our species to linger underwater for longer than ever before, that time was limited by the risk of "the bends." At the pressures of the depths, the nitrogen that a diver breathes in will dissolve into their blood and tissue. Normally this nitrogen is exhaled, but if the diver has been down too long, there is more nitrogen than the body can get rid of. On returning to the surface, this gas behaves like the spurt of bubbles that appears when you open a soda. The nitrogen ends up in places it's not supposed to be—causing decompression sickness, or the bends. This condition can lead to severe pain, stroke-like symptoms, and even death. Beyond around 100 feet (30 meters), a conventional scuba diver can only stay for a few minutes before having to decompress on the way to the surface, or risk the bends.

For many reasons, including the possibility of maintaining underwater assets and studying the effect of isolation, the US military began funding research to extend humanity's time underwater. In 1957, Navy physician George Bond discovered that breathing mostly helium, with a little bit of oxygen and a little bit of nitrogen, allowed humans to stay at the pressures of 200 feet (61 meters) in depth for more than a week. His human test subjects needed 27 hours in a decompression chamber to return to normal, but Bond's results suggested that humans could live in a deep seafloor structure at the same pressure as the water outside, allowing them to routinely swim in and out using scuba gear with no ill effects.

Soon after, two competing projects on the seafloor launched this idea into the public imagination. Starting September 6, 1962, diver Robert Stenuit spent 24 hours at 200 feet, breathing a helium-oxygen mix in a claustrophobic cylinder on the Mediterranean seafloor, for inventor Robert Link's Man-in-the-Sea project. Just four days later, Albert Falco and Claude Wesly lived for a week at 30 feet (10m) off France in the Conshelf I habitat, designed by Jacques Cousteau.

Cousteau and Link continued pushing into deeper waters. In 1963, a weeklong expedition in Conshelf II housed three aquanauts in the Red Sea at 30 feet (10m) and another two at a deeper 90-foot (27m) station, from which they dove as deep as 360 feet (110 meters). In 1965, six aquanauts lived in Conshelf III at a record depth of 330 feet (100 meters) for nearly 22 days in the Mediterranean. In 1964, Robert Stenuit and another diver returned to the seafloor at an astonishing 430 feet (131 meters) for Man-in-the-Sea II—though, due to technical failures, they stayed only 49 hours.

These successes invigorated US Navy interest. Under the medical supervision of George Bond (affectionately called "Pappa Topside"), more than thirty Navy divers lived on the seafloor during project Sealab, first in 1964 at 192 feet (58 meters) off Bermuda, and then in 1965 at 205 feet (62.5 meters) in the murky, cold waters of La Jolla, California. The team for Sealab II included three teams of ten humans and one non-human: Tuffy, a bottlenose dolphin. In addition to equipment tests and forty-six science experiments on the bottom, Sealab

II tested whether Tuffy could deliver messages and tools to the aquanauts, tasks at which he excelled.

The Navy planned for a third Sealab experiment in 1969, at a record-breaking 600 feet (182.9 meters). But during a repair dive, aquanaut Berry Cannon—a crewmember from Sealab II—died of carbon monoxide poisoning caused by a faulty dive canister. The Navy decided to abandon their efforts to house humans in the deep.

Academic and government research nonetheless continued in the decades that followed. Notable projects included Hydrolab, which hosted more than 700 aquanauts at 50 feet (15.25 meters) between 1966 and 1985; Aegir, at one point moored at a record-breaking 580 feet (176.8 meters) off Hawaii; Tektite, moored off the Virgin Islands at 43 feet (13 meters), which first modeled space missions for NASA astronauts and hosted the first all-female mission; and La Chalupa, where more than fifty scientists lived at 65 feet (19.8 meters) off Puerto Rico in eleven missions from 1972 to 1974. Other habitats were constructed worldwide, including in Japan, the Soviet Union, Great Britain, Canada, Israel, and Italy.

After 1970, interest in undersea habitats waned. They were extremely expensive, and contemporary technology left them dependent on a support ship. Crewed submersibles, ROVs, and AUVs began offering cheaper, easier, and safer access to the same data, as well as the ability to explore much larger areas.

However, deep-sea living hasn't ended entirely. NASA continues to send crews on simulated space missions to Aquarius Reef Base, built off Key Largo, Florida, in 1986, which is the only underwater habitat still operating. Meanwhile, Bond's dive physiology work pioneered an entire commercial industry. Saturation divers—who "saturate" tissues with a helium-oxygen mixture while being pressurized—became valuable to oil and gas interests, deploying for days to weeks at a time to maintain oil platforms as deep as 1,000 feet (305 meters).

Every handful of years, it seems, there is a news story about someone planning a new underwater habitat or hotel. None has yet come to fruition. Yet humanity's continued interest, despite the many ways we can now access the seafloor remotely, tells us something about our landlubber species' aquatic dreams.

FURTHER READING

Diver Alert Network. "Saturation Diving." https://dan.org/alert-diver/article/saturation -diving/.

Hellwarth, Ben. *Sealab: America's Forgotten Quest to Live and Work on the Ocean Floor.* Simon & Schuster, 2017.

Miller, James, and Ian Koblick. *Living and Working in the Sea.* Five Corners Publications, 1995.

United States Naval Undersea Museum. "Sealab II: Remembering the 'Tilton Hilton' 50 Years Later." https://navalunderseamuseum.org/sealab-ii-annivesary/.

motely operated vehicle (ROV) produced the first topographical map of the deep-sea rift where two continental plates were pulling apart. Towing instruments behind ships on sleds proved that technologies like cables and undersea cameras would be useful for true ROVs, small robots with thrusters that researchers on the surface could drive from the other end of a cable.

In the 1970s, the oil and gas industry began adopting ROVs to keep an undersea eye on their work sites; by the 1980s, ROVs were playing a role in huge discoveries, like the finding of the *Titanic*. Although these ROVs had to remain tethered to a ship, they could go deeper and stay longer than any crewed vehicle, and they could do so without the danger and cost of sending humans into the deep.

Around the same period, new technologies were coming online that pushed the door to the deep open even further. Cameras were beginning to go digital, making it possible to take still photos and record video footage underwater electronically, without large canisters of film that needed to be developed back in a photo lab. Ultimately this would lead to cameras capable of imaging in low light, at higher resolution, connected directly to increasingly sophisticated image processing. Computers, once machines the size of rooms, were shrinking rapidly until they could fit in a box that sat atop your desk—and marine engineers started putting primitive versions of these machines into underwater vehicles, with programming that meant those vehicles wouldn't need as much direction from the surface. Removing the need for a cable to send information to a vehicle had its benefits: To allow your vehicle to maneuver, you need a cable many times as long as the depth you're trying to reach, and that cable is heavy, can snag on underwater obstacles, and can be extremely dangerous if it comes under tension and snaps. If it breaks, that also leaves your vehicle stranded, with no power and no way to get back to the surface.

In the early Bluefin days, one of our first commercial customers was very skeptical of systems that were not attached to the ship, and did not believe the economic benefits. So we ran a comparison of tethered versus untethered for a survey being carried out in the Gulf of Mexico. While we were running the comparison, their team ran their tow vehicle into the seafloor and broke the tether. They were forced to mobilize a recovery vessel from shore to recover their very valuable hardware. No more conversation about risks of not having a tether!

Giving AUVs the power to think for themselves could also overcome some of the physical limitations that make communication underwater difficult; recall that only a limited range of sound frequencies travel far enough in water for a vehicle to receive them, and that there's a delay between when an order is issued and the vehicle reacts.

"In the ocean, you have pretty poor bandwidth, relatively speaking—like an old telephone modem," says David Mindell, a professor of aeronautics and astronautics at MIT, and the author of several books on robotics and autonomy. For an autonomous vehicle to really work, then, "the vehicle needs to know enough that you can give it high-level instructions, things that require on-board intelligence and processing," Mindell explains.

As computers advanced, that was finally becoming possible. Like ROVs, these new vehicles had the added bonus of not needing to keep humans comfortable and breathing in the inhospitable space below the waves—but unlike ROVs, computer-operated platforms would not need someone spending hours telling them what to do. In theory, such robots would therefore not only be much smaller than human-occupied vehicles like *Alvin*, but much cheaper to use for oceanographic research, given they wouldn't require long stretches of ship time to run nor long stretches of human time to operate.

There was another exciting aspect to giving our underwater proxies a "brain" of their own. Although this wasn't yet possible in the 1990s, already researchers were imagining that one day we could create vehicles smart enough that they could make decisions on their own, based on what they discovered out in the ocean. The potential of that was intoxicating. As Mindell puts it: "[For] most work in the science field, if you only program to look at what you're looking for, by definition you're never going to be surprised."

A few years prior, I finished my PhD in physics at MIT, after spending several years focused on creating superconducting sensors that could detect ultra-low magnetic fields. The only course I had taken related to ocean engineering during my education was in sonar design. However, part of my PhD work was funded by the MIT Sea Grant College Program, and from a distance, I had enviously tracked their efforts to start a program developing an underwater vehicle that could be fully autonomous, operating free of a tether and without human instruction. The directors of the program were excited by the technology; they had already supported the development of prototype digital acoustic communication systems and ROVs. However, AUVs were more of a reach, and finding a sponsor was challenging.

In 1989, fresh from graduating, I spotted a job ad on a leaflet in MIT's Infinite Corridor. MIT Sea Grant was looking for someone to run, and hopefully jump-start, their sputtering AUV development effort. In my visits to Sea Grant to report on our project, I had listened eagerly to the ups and downs of their efforts. When I saw the leaflet, I didn't hesitate.

They asked me to come by that afternoon. I knew the folks at MIT Sea Grant well; perhaps more importantly, they knew me. Much more than the average graduate student, I had been in and out of the Sea Grant office pitching ideas, reporting on progress, and learning about the various initiatives Sea Grant was funding. By the end of that interview, to my amazement and my wife's shock, I had a new job: starting and running the MIT Sea Grant AUV Laboratory.[9] With the vantage of time, it seems a dream job. But in the late 1980s, my friends thought I was crazy—what is an AUV anyway? Who needs one? And what did I know about underwater robots? But to me it was an opportunity to get in on the birth of a new and potentially very high-impact technology.

Our first vehicle was the *Sea Squirt*, a three-foot–long yellow cylinder that served as a test bed for the technology we would need to create functional AUVs. My expectation, going into the creation of *Sea Squirt*, was that programming it to carry out a mission without outside instruction would be the hardest part; after all, that was what made an autonomous robot unique. But giving our robot the ability to make basic decisions turned out to be one of the easier aspects of the project, even given the limited computer processors we had in the late 1980s and '90s. *Sea Squirt* operated on a type of computer programming called behavior-based architecture: Our team wrote code that dictated behaviors, each of which makes up basic parts of the robot's mission—things like "hold this speed and heading for this period of time," or "move between these depths," or the ever-important "do not run into the bottom." All of the mission's behaviors are then given priority levels, which combine to build individual tasks; all of those tasks combined to build fairly useful activities, such as "survey a grid of this size while moving in a zig-zag between the seafloor and the sea surface."

As it turned out, it was much more challenging to implement seemingly basic aspects of developing an ocean-going vehicle, such as communication or navigation. Once a vehicle is underwater, it cannot be reached by radio signals (or, later, GPS) to tell it where it is and how far it's moved. (Recall our poor petrel-beset *Odyssey,* incommunicado just by being inches beneath the sea surface.) Instead, we found, an AUV would need to compare how fast it had been going with how long it had been underwater to know when to change behavior or come to the surface. Power, too, is a constant problem for vehicles that are not "plugged in" to an electricity source the way an ROV is. It takes a whole lot of energy just to force something to move through the dense medium that is water, so propulsion ends up requiring most of any underwater vehicle's battery to this day. On top of that, at the time we were building *Sea Squirt,* the industrial electronics available at the time were extremely power-consumptive; from then on, many of our vehicles had to have custom parts built that were better adapted for the underwater environment.

Our early versions of *Sea Squirt* were indeed rough ones. For example, in attempting to save money, we first adapted model aircraft parts for some systems but found that they were far too unreliable. Still, it was a time of heady excitement. Unlike many engineers, who design to the most stringent requirements and then relax pieces as they go, we did the opposite, learning what could be built for cheap and then where we needed to make greater investments. To do this, I recruited an unusual team—unusual in that we had almost no ocean experience in the mix. After all, with no other AUVs like *Odyssey* or *Sea Squirt* in existence, we were writing the rules as we went along. While that meant we learned many lessons the hard way, it also allowed us to learn quickly and meant that the final product was shaped by the operational successes and failures of the real world rather than any outside expectations.

Great friendships are shaped by adversity, and some of my fondest memories come from this period. In the early years, the lab team drew heavily from the research community; at one time, four of us in the lab had PhDs in physics.

The final robot we produced operated only in sheltered environments, with sensors aboard that could measure oxygen concentration, water clarity, temperature and salinity in places like the Charles River, Boston Harbor, and local lakes and ponds. Missions lasted for a few hours at most. But *Sea Squirt* was a learning experience in figuring out the essentials that underpinned a working AUV. It allowed our lab to attract a great cohort of smart students with diverse experience, and it connected us with the researchers at other labs who would provide invaluable equipment and opportunities as we turned our sights to building the next AUV—one that might have true utility to ocean science.

That vehicle was *Odyssey*, and it benefited greatly from the lab's growing network and the burgeoning developments in the field. An agreement with Al Bradley from WHOI, who was developing electronics for their own AUV, the Autonomous Benthic Explorer (ABE), created a part pool for us to share: I provided the molds needed to create ABE's low-drag fairings, and he pitched in a motor controller for the thruster and actuators to control the motion of the robot's fins (like the flaps on an aircraft wing). We also further benefited from the work of lab member Cliff Goudey, who had preceded me at Sea Grant, running the fishery program. With the AUV program up and running, however, we quickly lured him to the dark side. Goudey had also worked on developing diver-powered submarines for underwater races; with his students, he'd used that knowledge to develop methods that maximized propulsion efficiency and minimized drag. The resulting *Odyssey* prototype was equipped to dive to 18,000 feet (6,000 meters)— although it never went below 6,600 feet (2,200 meters)—and weighed a bit under 350 pounds. It had a fiberglass body and sphere-shaped glass pressure vessels to keep sensitive equipment at atmospheric pressure when submerged; spheres are especially good at this because any point on the sphere is supported by curvature, like part of an arch, in two directions. And while *Odyssey* was originally intended to serve only as a prototype, our need to prove ourselves to sponsors pushed us on a more aggressive path—which, as described in the preface, led us to the Antarctic in 1992.

The base design of *Odyssey* carried our lab at MIT over the next eight years, and our decision to commit to seagoing field programs made us appealing to sponsors who might be otherwise wary of this nascent technology. Shortly after returning from our Antarctic cruise, an ONR oceanography program manager named Tom Curtin walked into the laboratory as a guest of ocean engineering faculty members. Curtin was then managing the ocean modeling and prediction program at ONR, and he was impressed enough by our results to fund us under the Sea Ice Mechanics Initiative (SIMI)—which, as the name suggests, aimed to figure out the physics of how ice floating on the sea surface moves and fractures. Our task was to map the underside of Arctic sea ice using an AUV. But to do that, we needed to build and test a new version of *Odyssey*.

The next version of the vehicle, *Odyssey II*, was a nearly complete redesign specifically made to work under ice. A key piece of that was installing a homing sonar on the AUV that communicated with a transponder at the ice hole, enabling it to find its way back to the place where we launched it. It navigated by listening to the pings sent out by beacons attached to the moving ice sheet,[10] which allowed the computer to calculate its speed and position. It also had a specially built system that allowed it to keep its heading, despite the fact that strong magnetic fields near the north pole could confuse the compass.

All this development occurred under tremendous time pressure. We were late additions to SIMI, and the field schedule and logistics were already locked in. By the time the funding came in, we had about six months to build this new vehicle from scratch, test it, and get to the Arctic. On the one hand,

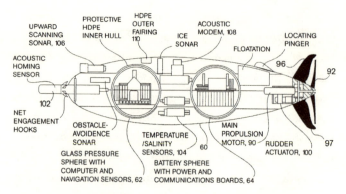

A line diagram from a patent Bellingham contributed to for the AUV *Odyssey II*, developed in the early 1990s for under-ice surveys in the Arctic. Its unusual feature is the acoustic array in the nose, surrounded by a ring of hooks. This was part of a homing system that communicated with and zeroed in on an acoustic beacon suspended under the ice hole in the middle of a net. The homing vehicle thus ran directly into the net, to which the vehicle's hooks became entangled, ensuring its safe retrieval.

the schedule was ridiculous, and it often seemed like there was no chance we could make it; plus, not a single person on our team had experience working at an ice camp. On the other hand, this was our big chance. If we failed, it would simply confirm to the world, and to ONR, that this technology was not ready for prime time. It was a period of long hours, intense pressure, and little sleep.

In early January, it became clear that we were not making fast enough progress, so I packed up the entire lab and headed north. We spent the next two months in the winter of 1993 sleeping in most of the rooms of a tiny one-star hotel on New

Hampshire's frozen Lake Winnipesaukee—rooms that were also converted to our electronics labs, when we weren't sleeping in a tent in the ice camp built a few hundred yards from shore. We "commuted" to our camp on rented snowmobiles (I particularly cherish a photo of Tom Consi, the biologist in our group, riding one of them while grinning like a speed demon). The original plan was that this would simply be a final test before deployment, but we were so far behind that we were still building the vehicle and writing the new code, testing incrementally as *Odyssey II* came together.

Early in our Winnipesaukee adventure, we were worried about all the gear we were leaving on the ice, and lab member John Leonard volunteered to sleep out in the tent. It was a miserable night for him: The air in the tent stratified, making it uncomfortably cold near the floor, with all the hot air near the top. Stresses built up in the ice relieved themselves in loud cracking reports that, at the least, were impossible to sleep through; Leonard spent most of that night wondering if his bed was about to plummet into the frigid waters below. After that, we decided we would bring as much of our gear as we could to shore each night and take our chances with the rest.

The long hours and minimal distractions greatly accelerated progress but at a personal cost that weighed differently on each of us; we had not planned to spend two months away from our respective homes before going on the deployment. Later on, this became the norm: The final crunch of getting ready to go to sea leads to long nights, lost weekends, and innumerable missed family events. But in winter of 1993, the team, and my wife, were reaching a breaking point, so I called a brief hiatus to repair relationships and recharge. I went home, although some stayed in Winnipesaukee and enjoyed the New Hampshire winter.

Gradually, all the bits and pieces came together. The first test ended up with *Odyssey* stuck in the mud; on the next trip out, we sent the vehicle out attached to a high-strength fishing line in case we needed to haul it back in. However, with progressive tests, we incrementally brought guidance systems and software online. The under-ice excursions became longer, and our operational roles and confidence grew. I'd like to say that the vehicle was completely ready for our under-ice mission, but one key piece of equipment from another lab had not arrived, so when we packed the vehicle and sent it to the Arctic, we added one additional piece of hardware: a mapping sonar, which we planned on completing in our month at the ice camp in Alaska.

Of course, the best-laid plans of mice and engineers alike rarely apply at the poles. To start, we walked into the barracks that passed for a hotel in Deadhorse, Alaska, shivering in −40 degree temperatures, wearing the clothes we'd worn when we took off in Boston. All of the cold-weather gear we'd ordered had been lost in shipping; the checked baggage with our backup gear also mysteriously vanished en route. Fortunately, the barracks were a short walk from the airport, so the trek was not far. On the way, the operator of some piece of oil equipment drove by us and offered us a ride. I remember the vehicle as having tires that towered over us, with a ladder you had to climb to the cab. As he leaned out through an open door over our head, we shouted up that we were almost to the hotel. He motored away, the machine's hulking silhouette looking like something out of a Star Wars movie as it faded into the blowing snow. Welcome to Alaska!

Deadhorse was, at that time, primarily an oil outpost, devoid of stores or restaurants. Once we realized that our shipment was missing, we separated into groups: some prepping for the coming deployment, and some scavenging for cold weather gear. Someone dredged up a jacket for me that was so heavy and stiff, it felt like someone had covered cardboard with cloth. It was ancient but indestructible, covered with assorted cuts and cigarette burns attesting to past abuse. I loved it, and even when we got our own stuff—which showed up just as we were leaving for the ice camp, adorned in the mismatched cold weather gear we had bought, begged for, and borrowed in town—I kept wearing that jacket.

A small Twin Otter airplane carried us to our ice camp on the Beaufort Sea, but the unexpected struck once again: We ended up staying only a week out of the four we had planned for testing. After we'd run just eight short under-ice missions, the weather took a turn for the worse, and the ice floe that was once solid ground began to break up beneath our feet. First, the runway that the Twin Otter would use on its return to pick us up broke off and drifted away leaving a widening gap of open water; a few days later, a crack began to open between the backup runway and the camp where we were sleeping. In the course of a few hours, we packed our gear and scurried over a makeshift wooden bridge across the gap, getting us to the plane and back to safety.

There were certainly successes that came from this truncated testing; our lab had proved itself as one that could confidently go to sea and get things done, and in all our tests in both Lake Winnipesaukee and the Beaufort Sea, *Odyssey II* never once missed its target when returning to the ice hole. But we were finding that the small vehicles we'd been focusing on didn't exactly match up with the vision of the people interested in AUVs. The Navy, for example, was envisioning much larger, more sophisticated platforms that could cooperate with their nuclear submarines—and given that the ONR was still one of the biggest funding sources out there for marine science, they often dictated the direction of the field. Yet we staunchly believed that the scientific community would not solely be able to rely on such large, expensive systems, which would not offer much of a difference from crewed vehicles like *Alvin*. An alternative vision was also needed.

At conferences and meetings with sponsors, I began pitching AUV work that would rely not on size but on numbers. The math behind battery efficiency works in your favor when you're running multiple vehicles, as does the cost balance. We had seen that these vehicles could carry mapping software that made them highly useful to military and research needs, but we had to figure out how to make those same parties want what we were offering.

In one conversation with Tom Curtin, I proposed a progressive series of AUV developments: First, we would test these vehicles by launching and recovering them off large ships; next, we'd see how a few of them did returning to a docking station after launch; then, eventually, we'd get to testing a large fleet of multiple AUVs, working in concert with little human intervention.

Tom thought about the idea for a few moments, and then, to my recollection, said something along the lines of: "Why bother with the first two? Let's go straight to the last."

Equipped with a new round of funding, the confidence from our field tests, and a feeling of vast potential, we were about to embark on something entirely new in the field: a vision of cooperative robotics that would change the way many of us saw the ocean—and revise my own perception of what marine robotics could offer in our quest to understand it.

CHAPTER 3

Teamwork

BY THE TURN OF THE TWENTIETH CENTURY, there was no question that ocean researchers and institutions found themselves in a vastly different place than they'd been at the start. Crewed vehicles and ROVs were now a common presence in both industry and research, and new technology was coming online seemingly every day. The launch of satellites that could gather information on the ocean and the atmosphere, starting with the short-lived Seasat in 1978 (which gathered only about forty-two hours of data),[1] was a watershed. It would take decades for the majority of satellite data to become widely available for science—high costs for access mean that, to this day, it's not attainable for everyone—but the potential of such technology was immediately clear.

Another huge leap forward came in the form of computer models. In the latter half of the twentieth century, scientists first grasped that there were immense processes at work in the ocean. Currents, temperature changes, and undersea "storms" behaved similarly to the constantly changing weather they saw in the atmosphere above us. By feeding measurements from the real ocean into our increasingly powerful computer systems, researchers could produce simulated oceans that acted like crystal balls, predicting what these enormous ocean weather systems would do next. These models would come to be an absolutely vital piece in demonstrating the potential of AUVs in oceanography. But symbiotically, AUVs would also be essential in helping these models realize their potential.

The early days of oceanographic computer models started with the usual big player: the Navy, which found itself grappling with an alarming change. In the early 1980s, Soviet submarines along the US East Coast, which had already been growing quieter and harder to track—the superpowers' arms race applied to stealth as well as weaponry—suddenly started disappearing altogether. The Navy's SOSUS listening technology, as well as its more active operations, simply couldn't find them, no matter how hard they searched.

There was no doubt that the subs were still there; after all, these were the Reagan years, when relations between the countries were at yet another all-time low. (Tom Clancy's famous novel and the subsequent film about submarine hunting, *The Hunt for Red October*, are based in this very time period.) Senior Navy officials began asking around at prestigious universities, hoping that some brilliant scientific mind might be able to explain where the vanishing subs might be hiding.

At Harvard University, they approached Allan Robinson, a geophysicist and expert in fluid dynamics. Robinson had a theory. He thought that the subs might be hiding within "warm core rings," water masses that broke off from the Gulf Stream to form independent little vortices of warm water. As Navy sonar operators had noticed on sunny afternoons decades earlier, sound moves more slowly through warm water, and so these warm whirlpools would refract sound away from the ring's center. Robinson thought that Soviet sub operators could effectively be using the eddies as sonar invisibility cloaks.

What's more, Robinson had a proposition for the Navy: He thought that with the right combination of computer programming and real-time data, he could predict the future of what the Gulf Stream was going to do next—and perhaps not just the Gulf Stream, but the entire ocean and even the planetary system that depended on it. To do so would have monumental consequences. The Navy was, naturally, interested in such a system for the power it would give to their anti-submarine efforts; knowing where Soviet subs might hide next would allow them to put their technological ears to the right patch of water. But scientists at the time were also becoming increasingly aware that the changing, swirling weather systems below the sea surface played a significant role in the ocean's biology, chemistry, and its ties to the planet's entire climate.

With just two weeks' notice, Robinson's grad students and postdocs were told that they wouldn't be going home for Thanksgiving of 1985: Instead, they'd be heading out on a Navy ship to monitor air-dropped bathythermographs (BTs) as they splashed into a meandering region of the Gulf Stream off Cape

Ocean Currents

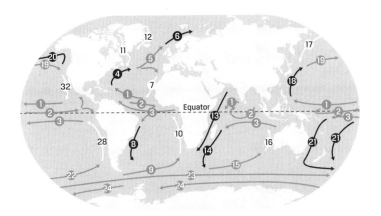

← Warm Current
⇐ Cold current
← General current

Currents:

1 North Equatorial
2 Equatorial Counter
3 South Equatorial
4 Gulf Stream
5 North Atlantic Drift
6 Norwegian
7 Canary
8 Brazilian
9 South Atlantic
10 Benguela
11 Labrador
12 East Greenland
13 Agulhas
14 Mozambique
15 South Indian
16 West Australian
17 Kamchatka/Oyashio
18 Kuroshio
19 North Pacific
20 Alaska
21 East Australian
22 South Pacific
23 Antarctic Circumpolar
24 Antarctic Subpolar

While we think about the Antarctic, Arctic, Atlantic, Indian, and Pacific Oceans as separate geographic entities, they are in fact ocean basins. All of these bodies of water are interconnected through currents.

Hatteras, North Carolina. Robinson and his students would return to this region five times over the next seven months to repeat the experiment, and with each visit, refine the system that they were building: a computer program called GULFCAST, which used satellite-gathered sea surface temperatures and the measurements from those dropped BTs to forecast how temperature fronts would move through the stream. These fronts might cause the stream to shift or straighten out, create meandering curves and loops, or snap off swirling eddies of warm or cold water along its path.

Over the course of those months, the model became powerful enough to go from forecasting a small portion of the Gulf Stream to predicting the ocean's weather across a span of more than three hundred square miles (about 777 square kilometers), stretching past the mid-Atlantic states and New England to the edges of Nova Scotia and Newfoundland.[2]

Robinson started publishing papers on the GULFCAST system in 1987, and its power was immediately apparent to the marine science community. Oceanographers began using GULFCAST to make predictions on how sound would travel over changing ocean conditions, while biologists developed a model of how changing ocean weather would shift the living communities within these underwater storms. An experiment called SYNOPS, a collaboration between the University of Rhode Island, University of North Carolina, and University of Miami studying the physics of the Gulf Stream[3], used it to decide where to sample next to get optimal data. NOAA was interested in using it to develop their own forecast system. And the Navy got what they'd been looking for out of their investment, using Robinson's model to create a Gulf Stream forecast that could predict sonar-reflecting features.[4]

The utility of predicting the ocean seemed obvious. But even as Robinson began turning his eyes toward a more ambitious application of his work—one that could forecast the ocean with even greater detail—he found himself wanting more. There simply wasn't enough data available to truly understand these systems in all of their complexity. Computers could describe how major currents like the Gulf Stream, or basin wide events like El Niño and La Niña, might evolve over months or years. But they couldn't say what the ocean in our own backyards might look like tomorrow, the way we might check the forecast to see if we need an umbrella before leaving the house. In 2003, I provided a quote to *The New York Times* that expressed my feelings on this problem: "The weather community is 20, 30 years ahead of us. And it's harder in the ocean."[5]

At ONR's ocean modeling and prediction program, Tom Curtin found himself frustrated with this very problem in the late 1990s. Although computing power was expanding rapidly as computers became more sophisticated, they were useless if they didn't have detailed information about the natural systems they were trying to predict. Plus, the information that Curtin's team could feed the models simply wasn't keeping up.

"The data to set up those models and then keep them on track was not expanding at all," he says. But it was clear to him that buying more ships, and sending them out to gather more data, wasn't the answer: "That's not going to be affordable, and the ocean's big—you could have hundreds of ships, but that wouldn't make much of a dent in sampling it."

Instead, Curtin and others in the oceanographic community were coming around to the idea of gathering data using many smaller vehicles, ideally ones that could be dropped off in the ocean and complete scientific missions on their own. Oceanographers had already seen the potential benefit of these kinds of tools on a temporal scale: In the 1960s and '70s, researchers had started dropping moorings into the ocean, where instruments could be affixed for months—and eventually, years—gathering information about that specific point.

As data from these moorings gathered, "you started to see things like eddies, waves you wouldn't have seen by just hanging sampling bottles over the side of the ships," says Curtin. But these stationary instruments had a downside: They couldn't capture ocean features that existed on large spatial scales, like weather fronts that might drift past a mooring in a matter of minutes or hours, or miss that spot altogether.

After our early work together in the Arctic, Curtin and I had conducted many a conversation on the utility of small AUVs working *en masse*—including the fateful conversation that ended the previous chapter. And I wasn't the only person he was talking to in this space. Curtin was not just a funder of other people's ideas; he was a visionary and a connector, the sort of person who could "see across the cubicles," as it were. He took all of these concepts that had been percolating in marine robotics and created a framework for it, called the Autonomous Ocean Sampling Network (AOSN). The vision was of a fleet of small, self-supported platforms, each with their own sampling skillset, working together to observe on-going ocean processes and feed the data they gathered into a real-time model.

Of course, many of the technologies that he foresaw as essential, like GPS and two-way satellite communication, were available only to large vessels. Nonetheless, Curtin captured funding from the Department of Defense in the mid-1990s to complete the idea, I pulled together a team to respond, and the AOSN Multidisciplinary University Research Initiative (MURI) was born. The program would go through many permutations over the next twelve years, but those first years brought together seven research institutions and countless talented individuals.

The AOSN project kicked off the busiest time of my life. As we took that project from an idea to a reality, my lab became a real seagoing enterprise. In those years, I spent as much as three months of the year preparing to go to sea, another three months actually at sea, and the remainder of the time traveling to Washington, DC, and visiting our various partner organizations. We operated AUVs over the Juan de Fuca Ridge off Washington State, off Bermuda, off New Zealand, in the Mediterranean, and for two of the most challenging winters in my life, in the freezing Labrador Sea off Newfoundland. With every trip out, going to sea began to feel second nature. When I first went to sea in Antarctica, it felt like going to a fancy event where I didn't know what piece of cutlery to use; with enough practice, I became comfortable enough in the field to manage others who were having the experience for the first time.

As demand for our systems grew, I cofounded a new company, called Bluefin Robotics, to spin off some of the AUVs we had developed at MIT for sale to commercial and military interests. I was juggling the directorship at Bluefin even as I was approached by a familiar colleague, Marcia McNutt, to join her on the West Coast at the Monterey Bay Aquarium Research Institution (MBARI). It was a dream job, so I packed up my family and moved west. I barely had my feet under me at MBARI when Tom Curtin approached me to ask if I would become principal investigator on the second round of the AOSN MURI project. I couldn't help but say yes. It felt like a chance to clean up some unfinished business; we had demonstrated that our ideas for autonomous technology worked, sure, but we had never created a program that could truly demonstrate its power. This was that chance.

By the time the first phase of AOSN MURI concluded in 2000, the project had developed most of the technological infrastructure needed to make our original vision possible. One of the major successes of the program was the creation of three new AUVs—the Webb Slocum, the SeaGlider, and the Spray—that used gliding, rather than propulsion, to move through the water. The glider concept came from oceanographer Doug Webb, who had introduced it to the community through a science fiction narrative written by Henry Stommel, but based on Doug's ideas, in 1989.[6] Tom recruited Webb to AOSN to make the idea into reality. Glider AUVs work on the same concept an airplane uses to fly, generating lift over its wings, which moves the vehicle forward. While slower than a vehicle with a propeller, the lack of propulsion means that gliders can stay at sea for long periods and travel long distances, all without the need for humans to change batteries or service an engine.

Our fieldwork in the 1990s, however, had mostly focused on making sure that all the hardware was working at sea. We didn't have the infrastructure in place to actually do more than the bare minimum with the data that the vehicles were gathering; we could barely even use it to adapt the plans for our next mission.

In 2003, a new phase of the project dubbed AOSN II provided a true demonstration of the power of these systems, and of the power of the data that they collected. Over a month-long period, we consistently kept eleven vehicles at sea. Most of these platforms were gliders, but they were complemented by a few faster, more sophisticated propeller-driven AUVs. The gliders alone took over 13,000 measurements of salinity, temperature, and depth over that month. A low-flying aircraft, gathering parameters like air temperature, atmospheric pressure, wind speed and direction, and atmospheric aerosol count, provided some of the information that satellite systems (not yet readily available to us in 2003) would later fill in.

This was the vision of ocean science that I had been imagining for so long: fleets of AUVs working in concert with shore-based scientists, making it possible for us to collect unprecedented amounts of information on the ocean's workings, and saving the scientists onshore both time and money. I have a distinct memory of sliding into a booth for lunch with some fellow project members at Phil's Fish Market and Eatery in Moss Landing, California, on a spring afternoon in 2003. I looked over the sand dune in the direction of the shimmering Pacific, where just a few miles away, our AUVs were gliding or propelling their way at various depths through the waters of Monterey Bay. "We could get a beer if we wanted, and we're still doing oceanography," I remember saying. At the time, I was certain that it wouldn't be long before every university, every research institution, had a fleet of AUVs of their own.

When it became apparent that our equipment was actually working, toward the end of the first round of AOSN, Tom Curtin brought in Allan Robinson, predicting—correctly—that our AUVs had the power to offer the abundant, widespread real-time data that his predictive models had been missing. Robinson and his modeling group started putting that data to work in the Harvard Ocean Prediction System (HOPS), which they had developed in the years after their work on GULFCAST.

The HOPS system and two complementary models, the Jet Propulsion Laboratory implementation of the Regional Ocean Modeling System (JPL/ROMS) and the Navy Coastal Ocean Model/Innovative Coastal Ocean Observing Network (NCOM/ICON), gobbled up the data gathered by our AUVs and spit out "nowcasts." These nowcasts processed the data in hours, rather than the months to years that it had once taken to go through all of it by hand, identifying oceanographic patterns like currents, fronts, and eddies. This allowed us to shift our plans for the next mission for the vehicles, a strategy called adaptive sampling. Working through a shared computer portal, all the teams involved would then remotely vote on what sampling plan seemed like the best use of our resources. This sort of remote collaboration might seem trivial now, but it was revolutionary at the time.

What's more, the computer models were becoming powerful enough to actually predict the future. They used the vehicles' data to predict the next day's pattern of coastal upwelling, the rush of cold, nutrient-rich waters from the deep to the ocean surface. (Upwelling plays a major role in marine ecosystems, feeding the plant-like phytoplankton that form the base of the ocean's food chain; as a result of this upwelling, the area around Monterey Bay is rich with marine life, from multitudes of sardines to flocks of diving birds up to seals, dolphins, and whales.) After observing how well those forecasts matched what actually happened, the modelers would then adjust their computer programs, training the models day by day until they could paint a picture of the next day's view of the ocean with remarkable accuracy.[7]

Additionally, by shifting the relative importance of winds and currents in those models, we could also see how they affected factors like ocean mixing. Ocean mixing, in turn, fueled or inhibited phytoplankton blooms, which could help predict the boom or bust of fisheries that relied on phytoplankton for food. We were beginning to see the ocean in new dimensions, highlighting connections that had always been otherwise invisible.

The AOSN II field deployments succeeded beyond even our own expectations. Simultaneously, people outside of the project were starting to take note of what it was accomplishing. By the time phase II started, a number of other institutions had already started to build their own fleets of gliders for sampling. As the project progressed, more and more people started joining in on the experiment planning sessions; by the final meetings, the room was packed with people—some of whom flew in specifically for these sessions—and another fifty or so joined us by teleconference.

And technologically, we had succeeded in taking a great leap forward. The observations we gathered would not have been possible with the moored technology available before it; nor would they have been so successful without the level of autonomy we achieved in our AUV fleets.

"AOSN put us on another level," Curtin recalls today. "There hasn't been anything done like it since—and I'm amazed every time people still working in this space tell me the same thing."

In the decades that followed, we've seen the impact of AOSN slowly radiate outward through science at large. Our Princeton University collaborators used the experiments as a jumping-off point to launch the Adaptive Sampling and Prediction (ASAP) program, which explored cooperative strategies for improving the observational power of teams of autonomous vehicles and other mobile platforms. The Navy used it to develop Persistent Littoral Undersea Surveillance Network (PLUSNet), a framework putting AUVs to work protecting coastlines from foreign submarine threats. The Naval Oceanographic Office also operates a vast network of gliders that is constantly collecting data from all over the world.[8]

My vision of a fleet of AUVs for every institution, accessible to any research problem, hasn't yet come to fruition. But a different vision has come into existence instead. Regardless of how autonomous they might be, these multiplatform systems are still expensive to start up and to maintain, both in dollars and in human work hours. As a result, multiplatform systems in place today are most often run by large government organizations, and focus on constantly collecting basic ocean data that scientists can access for their work. The NOAA-led

Integrated Ocean Observing (IOO) system uses an array of moored buoys, water gauges, radar installations and gliders to collect parameters like temperature, salinity, and pressure, as well as to carry out some specialized functions like tracking tagged marine life and collecting data on man-made sound underwater.[9] The other system, which stands as perhaps the most ambitious project of this kind, is the Ocean Observatories Initiative (OOI). Funded by the National Science Foundation, OOI extends to all of the world's major oceans, consisting of a series of "arrays"—regional collections of cabled observation stations and autonomous vehicles, each with as many as forty-five different instruments that can collect data on two hundred different ocean parameters.[10] Both of these efforts, it's worth noting, also use gliders that we developed during AOSN.

All of these systems work together to collect data that play an unseen, but often surprising, role in our daily lives. Understanding temperature changes and the churning of ocean currents allows for better weather predictions here on land. Having lots of sensors out there helps scientists keep track of phytoplankton blooms out at sea, where they're largely invisible but can play a huge role in fueling fisheries that in turn feed humanity. And on an even bigger scale, better data about the physical and chemical processes in the ocean gives us a better understanding of how it's taking up carbon and absorbing and transporting heat, thus letting us feed more accurate data into the climate models that give us a sense of our potential futures.

* * *

In addition to these large-scale, often ocean basin–sized efforts, we've also seen real-world examples of how transformative it can be when AUVs cooperate on a smaller scale, focused on a singular project. One of the best examples of this is the search for Malaysian Airlines Flight 370, which I highlighted in chapter 1.

When we left the search, it had been more than a month since the plane crashed. Bluefin's AUV *Artemis* was deployed to the search area, a desperate stab in the dark as the clock ticked well past the expiration date for the battery of the downed aircraft's flight recorder.

Despite the surprise of finding out that the seafloor was deeper than shown on maps, some quick calculations by Bluefin's engineering team found that *Artemis* should be capable of withstanding the pressures of a dive deeper than its pressure rating. With bated breath, the crew sent her down to scan the bottom. *Artemis* was using a type of sonar called side-scan sonar to scan the Indian Ocean's craggy bottom, sending out waves of low-frequency sound in a fan shape. This creates "slices" of the seafloor that can be compiled into an image with nearly as much detail as a photograph. *Artemis* survived the trip below 4,500 meters, bringing back detailed scans of the seafloor below, but it, too, saw nothing that looked like a Boeing 777. Faced with poor maps and a rapidly spreading debris field, it was becoming quickly apparent that finding the lost plane would require much more than the effort of a single vessel.

The search for MH370 concluded in January 2017, just under three years after the plane vanished. By that time, the final effort looped in a whole fleet of different vessels: twenty ships, twenty-one aircraft, one submarine, more than eighty sonar-equipped sonobuoys, and five underwater vehicles, including two different towed instrument platforms (commonly called towfish), one tethered ROV, and two different types of AUVs working in concert with their surface ship.[11] It used multiple types of sonar: passive sonar from the sonobuoys; side-scan sonar from *Artemis;* ship-based multibeam sonar, used to roughly map the bottom; and synthetic aperture sonar from the AUV *Hugin,* which bounces sound off the same feature from multiple angles in order to create a highly detailed image.[12] The early search was conducted by a fleet of ships and one AUV; by the end of the search, a single ship was tending eight AUVs. Not a single one of these methods yielded evidence of MH370—not even a fragment of the plane showed up on the seafloor imagery. Even in the face of some of mankind's most advanced technology, the ocean kept her secrets.

The disappearance of MH370 was a tragedy: for the people on board, for their families, for the international community. Yet just as the world wars led to a wholly new understanding of how our oceans work, the search effort demonstrated how much we can learn about the oceans when we apply the right tools. And because it was done entirely with tools that we had in hand—no new technologies were invented for this search— it underlined that achieving this level of scientific accomplishment has always been possible, and need not come from an event with such a devastating cost in human lives.

The MH370 search transformed this part of the Indian Ocean from nearly unknown to one of the best-mapped seafloors on the planet. The search area was one of the largest underwater surveys ever conducted, covering an area of more than 274,000 square miles (440,960 square kilometers) and collecting more than 107,000 square miles (172,200 square kilometers) of bathymetric data, revealing the depths and features of the bottom.[13] The resulting maps had a resolution at least fifteen times higher than what existed before. The sonar scans revealed previously unknown seamounts rising from the seafloor, illuminated the details of Mount Everest–sized mountains and deep subsea canyons, and provided geologists with new details about the breakup of tectonic plates here 40 million years prior. Viewing the before-and-after maps is like comparing Vincent van Gogh's *Starry Night* to a space telescope's images of the universe.

These detailed maps can now be used to seek out new fishing grounds, predict how tsunamis could move through the area, and model the effects of climate change. Such a prolific output demonstrates just how much information can be gained about the ocean when marine science deploys its available technology in concert, using each to fill in what the others lack.

This type of multiplatform cooperation is gaining momentum in science, but it's still relatively rare. The cost of sending a ship full of humans to sea, and all the fuel needed to keep both running, is still immense: The search for MH370, one of the most expensive search-and-rescue operations in history, came out to around US $130 million. Even ordinary ship operations have a high price; the UNOLS (University-National Oceanographic Laboratory System) fleet, the set of ships available for academic research through US federal agencies, have total operating costs of around $100 million per year.[14]

Funding, therefore, stands as the primary obstacle to implementing this type of cooperation for scientific research, as well as to search efforts like the one for MH370. Yet our field continues to envision how that might change. Deploying multiple underwater vehicles, working in concert, can lower the cost of an operation; when working together, individual vehicles can move more slowly even as the cumulative system achieves the survey faster, using less energy and requiring fewer expensive batteries. With every vehicle you add to this hypothetical survey, it becomes more cost- and energy-efficient. Furthermore, each of these vehicles can carry increasingly sophisticated and capable payloads, dramatically increasing the amount of information they can gather.

Most marine researchers don't have access to supernumerary fleets of autonomous ocean vehicles that can do most of their data-gathering for them. Nonetheless, AUV fleets are growing, and some projects are collecting an extraordinary amount of data. As with many things in ocean science, progress ebbs like a rising tide; it often has to pull back before creeping to a new high. I get the feeling that today, the tide is on its way back up.

CHAPTER 4

The Living Ocean

BY 2006, I'D BEEN WORKING IN OCEAN SCIENCE for more than two decades. I lived in central California, next to one of the most productive patches of ocean in the world, and on any given day I could drive down to the coast and be lucky enough to see egrets and cormorants soaring overhead, sea otters cracking shellfish on their bellies, gray whales poking their heads up from the ocean's surface, and common dolphins threading through its waves. But for a long time, all of this life meant nothing to my career: If the ocean's waters were completely devoid of life, my work would have been the same, as far as my vehicles were concerned.

During the course of AOSN and my first few years at MBARI, however, that started to change. I had many conversations with people like microbial genomicist Ed DeLong, biological oceanographer Francisco Chavez, and molecular biologist Chris Scholin. The picture that emerged from these conversations was of a living ocean, an entity composed of microscopic organisms, rather than just a medium in which organisms live.

The liquid our vehicles swam through was not just water molecules and salt ions; it was packed to the proverbial gills with microscopic life, and if all of that life were to die, the Earth—and indeed, many of the parameters that I was training autonomous vehicles to follow—would be fundamentally transformed, for the worse. Soon microbial ocean life was all I could talk about, a fact that likely drove the marine biologists in my radius crazy; this was not exactly a new fact for them.

In the 1980s, microbial biologists developed a new technique called epifluorescence microscopy, in which they dropped special dyes into seawater that attached themselves to the DNA and RNA of any cell within. When viewed through a microscope under fluorescent light, those dyes glowed, enabling scientists to get a count of just how many cells could be found in a single milliliter of seawater.

This discovery utterly revolutionized our understanding of life in the sea. Previously, the only way to guess at the abundance of microscopic life in the ocean was to put seawater on a petri dish and see how many colonies arose from each sample. But as researchers later learned, many ocean microbes are difficult or impossible to culture on a plate. Petri dish estimates had put microbe abundance at around 100 microbes per milliliter of seawater; for comparison, this is like scattering a handful of poppy seeds in an Olympic-size swimming pool. With the help of epifluorescence microscopy, researchers upgraded that estimate to a *million* microbes per milliliter of seawater—and that was just the bacteria. These bacteria lived in a miniature world of their own, filled with marine viruses, single-celled Protozoa, tiny animals called zooplankton, and plant-like algae. Every drop of ocean water was squirming with creatures being born and dying, reproducing, eating, squabbling over resources, and producing many of the things that keep the rest of the planet functioning.

"Previously, we didn't think there was anything living in ocean water smaller than a micron," says Sallie "Penny" Chisholm, an MIT biologist who has dedicated her research to *Prochlorococcus,* the smallest and most abundant family of photosynthesizing bacteria (called cyanobacteria, or blue-green algae) in the world's oceans. She recalls that during her postdoc at Scripps, her advisor was working on what he called "the wee things" that could pass through one-micron filters. "Everybody thought it was broken-up big cells, like chloroplasts. But I'm sure it was *Prochlorococcus* and other things like it."

Chisholm first noticed these "bugs" when she and her postdoc, Robert Olson, tried running seawater through a flow cytometer, a device commonly used in medical applications that sorts individual cells from a sample using a laser. Olson noticed that the laser stimulated minuscule flashes of red fluorescence from the seawater—a signal that they thought might be coming from life but might also just be electronic noise. It wasn't until further work photographed tiny, fuzzy green cells under an electron microscope that Chisholm and Olson's hunch was confirmed: These were living creatures, as small as a half-micron wide. You could arrange 38,000 of them in a line and they would fit across the face of a US penny.

"The technology opened up this whole different realm, made us realize that we didn't even know what was going on out there," Chisholm says. They called the new organism *Prochlorococcus,* the "primitive green berry."[1]

Around the same time, WHOI biologist John Waterbury, one of the first scientists to apply epifluorescence microscopy to ocean microbes, was using this technique to study bacteria in the Arabian Sea. He noticed that some samples would glow brilliant orange under fluorescent light even before any dye was added to them. After a year of work, he and his colleagues were able to culture the culprit in the lab, a cyanobacterium that would be called *Synechococcus.* Together, *Prochlorococcus* and *Synechococcus* are not just the two most abundant photosynthesizers in our oceans, but they may be the most abundant organisms on our planet.[2] These two families of cyanobacteria create a huge portion of the oxygen we breathe every day— especially *Prochlorococcus,* which is an oxygen workhorse. Estimates vary, but it's thought that these green berries produce at least 13 percent, and perhaps as much as 48 percent, of photosynthesis-made oxygen in our atmosphere.

The role of other microscopic species can't be downplayed, either. Zooplankton and phytoplankton form the base of the marine food web, providing food for everything from brine shrimp (some of which are popularly known as sea monkeys) up to whales. Viruses keep populations in check, and they can actually transfer genes between microbial species, helping to spur evolution.[3] Many forms of bacteria perform reactions that move elements around within the ecosystem, providing other creatures with essential nutrients like nitrogen; as DeLong puts it, in the ocean, "microscopic life *is* biochemistry."

But even as all of these technologies made it possible to see these tiny creatures in the laboratory, the platforms we had for observing them in the wild weren't doing the job.

Chris Scholin, the former president and CEO of MBARI, was a graduate student working on DNA sequencing ocean microbes at WHOI in the early 1990s. He remembers heading out on a dilapidated lobster boat to collect samples off the New Hampshire coast alongside a group of physical oceanographers.

"We'd come back in, and the guys doing the physical oceanography, they had all their data before we even stepped off the boat—while those of us getting the cell counts would just have jugs of water," Scholin remembers. "If we were lucky, a week later we would know that the things we were interested in were at this place at this time. But by the time you know that, everything is different."

As we had seen with other ocean processes, a stationary mooring or a sample from a ship could only provide data illuminated in a spotlight of space and time. Yet biological systems are often patchy and temporary. They can easily be split apart or whisked to a new place by changing ocean conditions, taking them out of view of a stationary mooring. As Scholin knew all too well, they can bloom, fade, and die in the time it takes for a ship to get back to land and process a sample. In order to truly understand marine organisms, we had to follow them.

Fortunately, in 2006 I had just finished a stint as MBARI's Director of Engineering and was in need of a new project. After a series of conversations with Francisco Chavez about biochemical observation systems, and with the support and interest of other MBARI scientists, we proposed merging the capacities my team had gained from AOSN with Chavez's longtime interest in observing the shifting nature of coastal ecosystems.

In addition to providing completely novel insight into how ocean biology worked, many of my collaborators—including Scholin, who came to MBARI in 1992—also saw a practical use for such a system. As our oceans warm, coasts are seeing more and more frequent harmful algal blooms (HABs). These blooms can have profound health effects on marine life and humans alike, ranging from breathing problems to neurological issues all the way up to death. If we could have some advanced knowledge of these blooms, it might be possible to warn people to stay out of the water and refrain from eating shellfish—which concentrate toxins from a HAB as they filter the algae from the water—before anyone became sick.

Coupled to biological applications from people like Chris Scholin was a new low-power vehicle concept that I had been

Harmful Algal Blooms and Climate Change

It starts with something very, very small: a single-celled organism floating in the ocean that turns sunlight into food. When times are good—lots of nutrients, comfortably warm water—this tiny cell multiplies. When times are *too* good, one cell can become billions; in the space of days, this tiny thing can shut down local fisheries, send marine mammals washing onshore in distress, and leave human beings hospitalized and even dead.

These tiny creatures are algae, plantlike cells that form the foundation of the marine food chain. But in recent years, coastal waters around the world have seen algal blooms of unprecedented size and duration, with devastating effects for local ecosystems. The culprit? Scientists suspect climate change, although finding a single cause is not quite so simple.

Algal blooms pose two potential threats to coastal ecosystems. One is simply part of the circle of life: Like everything else on Earth, when algae die, bacteria eat them. Those decomposing bacteria use oxygen as they work, and when algal blooms are especially large, bacteria can strip most or all of the oxygen from the water around them. This makes areas called "dead zones," which can kill any animals or plants unlucky enough to get caught inside.

The other threat from algal blooms comes from species that naturally produce toxins. Some of these algae have direct impacts, such as *Karenia brevis*, which causes the rusty-colored waters known as red tide. *K. brevis* also releases a toxin called brevetoxin, which can cause respiratory issues and skin irritation. Other impacts are indirect. As shellfish, like oysters and clams, filter toxin-producing algae out of the water, those toxins concentrate in their flesh, harming anything that eats them. This mechanism was responsible for one of the most famous harmful algal blooms in history, when in 1987 four people died and more than a hundred fell ill after eating mussels from Prince Edward Island. These mussels had been contaminated by domoic acid, a toxin produced by a species of algae called *Pseudo-nitzschia* that causes intense gastrointestinal issues as well as neurologic problems. One of those problems includes short-term memory loss, which gave this condition the name "amnesic shellfish poisoning."

Researchers have known for a long time that adding lots of nutrients, like nitrogen and phosphorus, to water can kick-start algal growth; this link was established in freshwater in the 1970s, when phosphorus was traced to huge blooms on Lake Erie. But most algae species do better in warm water, so researchers have long suspected that ocean warming could also play a role. In 2017, a study in the *Proceedings of the National Academy of Sciences (PNAS)* used computer modeling to show, for the first time, that rising water temperatures could be blamed for algal blooms broadly—across entire oceans, not just in individual ecosystems. These blooms are getting bigger and more common: A 2023 study in *Nature*, based on satellite imagery, found that between 2003 and 2020, algal blooms worldwide became 59 percent more frequent and grew 13 percent bigger.

Yet the climate link is also complicated. The *Nature* study found that blooms in high-latitude ecosystems, like around Alaska and in the north Pacific, were particularly influenced by ocean warming, and that changing circulation patterns caused by climate change made algal blooms more common east of South America and along the Canary Islands.

In contrast, the climate influence made blooms less common in some places—like in the California Current—and in others, additional factors play a bigger role. Around China and in the Arabian Sea, more fertilizer seems to be fueling blooms, while bigger aquaculture operations in Finland, Algeria, and Argentina may be responsible for algae there. However, that doesn't mean the climate isn't playing a role in these places, too. Climate change is often referred to as a threat multiplier: Even if it does not directly cause a problem, it can boost the effect of other factors, making issues worse than they would have been otherwise. This certainly seems to be true in the complex equation that is our ocean, where a changing climate is giving these blooms the boost they need to end up on top.

FURTHER READING (AND LISTENING)

Dai, Yanhui, et al. "Coastal Phytoplankton Blooms Expand and Intensify in the 21st Century." *Nature* 615 (2023): 280–284. https://doi.org/10.1038/s41586-023-05760-y.

Gobler, Christopher, et al. "Ocean Warming Since 1982 Has Expanded the Niche of Toxic Algal Blooms in the North Atlantic and North Pacific Oceans." *PNAS* 114, no. 19 (2017): 4975–4980. https://doi.org/10.1073/pnas.1619575114.

"What Connects Bones, Bird Poop, and Toxic Green Slime? Hint: Without It, Half of Us Wouldn't Be Alive Today." *Gastropod Podcast* (2023). https://gastropod.com/what-connects-bones-bird-poop-and-toxic-green-slime-hint-without-it-half-of-us-wouldnt-be-alive-today/.

exploring, drawing on lessons learned from AOSN. At the time, I was having conversations with colleagues about the relative strengths and weaknesses of gliders and AUVs, and where these different designs would be most useful. Off the coast of the United States, where vehicles could in theory be launched easily and readily, it looked increasingly to me like the solution was to build a system with the best attributes of an AUV *and* a glider.

But the real impetus came from a little battery technology company near Berkeley called PolyPlus Technologies. They had an experimental battery with an energy density perhaps three times the best we had available—three times more power for the same size battery. The only hitch was that it operated at very low power levels. If I could figure out how to make a vehicle operate at those low power levels, we could build a small vehicle capable of crossing the Pacific.

As I worked through the calculations, a vehicle began to take shape: one capable of going fast when needed, but capable of dropping to progressively lower power modes when you wanted to draw out your endurance. The results, frankly, looked too good to be true. Indeed, even if the new batteries didn't work out, the vehicle was still worth attempting—if MBARI was willing to let me build it. At that time Marcia McNutt, who invited me to Antarctica back in 1992, was MBARI's president and CEO. Luckily, she said yes.

The vehicle project was embraced by colleagues, and together we began working on a biological observing system that would do for the living ocean what AOSN did for the physical ocean. This project became known as CANON: the Controlled, Agile, and Novel Observing Network Initiative. While AOSN provided essential lessons for CANON, from the start it was clear that this was a different beast. AOSN had focused on the physical ocean, and there were known, solid mathematical properties we knew governed the physical ocean: most notably, the Navier-Stokes Equation, which describes how fluids move, are compressed, and change with temperature and pressure. Once AOSN's vehicles were in good working order, it hadn't been a huge leap to feed our data into mathematical models built around those equations, and use them to predict the future.

But no such equations existed to predict the biological ocean. Even today, this type of science is in a much earlier stage. It's like pre-Newton astronomy: We can see the patterns, but we still have not discovered gravity. And in contrast to AOSN, which was about enabling prediction, this new venture would be about understanding the fundamentals of marine microorganisms in the ocean: where they live, how they respond to the environment, why they die.

To do this, we would need a totally different type of AUV. That vehicle would need to be capable of traveling all the way out to the productive waters of the California Current and back, a span of more than five hundred miles (roughly 800 kilometers), while still having enough power to do useful science in the current system. Its batteries needed to last long enough to monitor algae from "bloom to bust," which could be up to four weeks. It needed to be able to carry sensors that captured biological and chemical properties, like chlorophyll levels and cell counts, as well as the typical physical oceanography sensors we'd need to connect blooms with broader ocean conditions, like temperature and salinity. And the vehicle needed to be affordable enough that we could make and operate a bunch of them simultaneously, which meant that it needed to be relatively small.

The result, after two years of development, was the *Tethys* long-range AUV (LRAUV). It weighs in at 260 pounds (118 kilograms) and just about 7.5 feet long (2.25 meters), and operates with power management as a central goal, with sensors that operate intermittently or turn off altogether for periods in order to reduce consumption. It also has the ability to operate at different speeds: one meter per second for most data-gathering; 0.5 or 0.7 meters per second to maximize range; or simply drifting, propellers off but holding a constant depth, allowing it to gather data while moving with a water mass.[4] *Tethys* vehicles—today, there are around a dozen of them in operation—now have an estimated range of 1,200 miles (about 1,900 kilometers) running at full speed. Although it would take a long time, our calculations estimate that one could make it from Moss Landing in California all the way to Honolulu running in power-conserving mode.

Our first attempt to monitor phytoplankton using *Tethys* earned the nickname of the "robo-ballet," as three vehicles descended on the bloom: a surface drifter to provide a location reference, the *Tethys* to map the plankton patch around it, and another AUV called the *Dorado* to periodically sample nearby. These tests almost immediately led to upgrades. We gave the *Tethys* the ability to use chlorophyll fluorescence to autonomously track the phytoplankton bloom, which meant that it didn't need to surface and check in with shore to get new directions. We also gave the vehicles the ability to track each other. While the *Tethys* knew where the phytoplankton bloom was, it had no actual idea where *it* was; unable to connect to GPS below the surface, a vehicle doesn't have any reference points for its own position on Earth. The drifting surface vehicle gave both the underwater system and us humans back on land an idea of where a bloom was taking place and served as a communications gateway through which we could relay orders for the fleet's next moves.

By the time we were developing and advancing *Tethys*, Chris Scholin was almost a decade into developing a device that would become another powerful tool in the LRAUV's chest: the Environmental Sample Processor, or ESP. This system is among the first capable of actually processing samples on-site, allowing researchers to collect a constant stream of real-time information about the water passing by. "If you think about the weather, we have all these automated stations all over the place," Scholin explains. "That was the concept, really: collect the data at their spot, put it together, and you get an incredible view of what's happening in the environment."

The first ESP focused on identifying toxins from harmful algal blooms passing by, but today it can do even more. These devices can also analyze environmental DNA (eDNA), the tiny genetic signatures that all living creatures leave behind in the water as they eat, excrete, reproduce, and shed their skin. "We can see fish migrations, invasive species, pathogens—waves of organisms," Scholin says. "It's pretty profound. It's like a genomic weather map."

While the first ESP was the size of an oil drum, the latest models are small enough to fit on an AUV, allowing researchers to truly combine biological signals with oceanographic data about the environment that those creatures live in.

Our power to understand the ocean in all of its complexity, both living and nonliving, has increased dramatically over a very short period, thanks to the reciprocity that allows science and technology to enable each other. Of course, I remind myself constantly that the strength of this technology is entirely dictated by how we choose to deploy it. Marcia McNutt, looking back at her career in enabling technological exploration not just at MBARI but across the sciences, explains this best: "I was always struck by the fact that any time you took new sensors, new eyes, new ways to explore the ocean, out into the field, you discovered things that you never even knew existed," she says. "It wasn't just a tool that would complement other things you were doing; it would actually open up whole new vistas of exploration."

My favorite example of this comes from the time that we chose to put a camera on a *Tethys* LRAUV while working on a moored docking system for the vehicle. When my team reviewed the footage, we were shocked to find ourselves looking at waves upon waves of undulating sea nettles, pale pink or orange jellyfish with a cape of frilly, stinging tentacles. There had to be hundreds of them passing by the camera as the *Tethys* puttered through their midst.

As I thought about these nettles, suddenly a pattern slotted into place. During previous experiments, we had noticed that sometimes, oceanographic conditions would be absolutely perfect for phytoplankton blooms, but our vehicles would detect no signal of them in the water. Watching the hypnotic progression of jellyfish pass *Tethys*'s camera, we realized that this could be the culprit: massive packs of jellyfish that were gobbling up the phytoplankton before we humans had a chance to ever see their signature in water samples.

Yet these were the days before the ESP's environmental DNA sampler was small enough to fit on an AUV, and nothing aboard the robot was equipped to detect jellyfish. Without the camera, we would never even have known that they were there. By simply changing our frame of reference on the ocean, we had provided new clues to a long-standing question and discovered new dimensions to this space—a space forever reminding us of how much more we have to learn.

CHAPTER 5

The Ocean in Our Kitchens

IN THE PREFACE, I wrote that the ocean may seem to many land dwellers like a distant, almost alien world. For most of our species' history, that was true of the bulk of humanity: Even while some adventurers were crossing oceans, the average person—unless they lived right on the coast—probably felt that the ocean made no difference to the food that they ate, the growth of their crops, the health of their children, or the weather that passed by every day. And in some ways, this is still the case; when the ocean enters the public's consciousness nowadays, it's usually only due to some disaster, like a shark attack or the sinking of a vessel.

The difference today, I think, is that few people can claim ignorance of the ocean's importance. With the deluge of regular news stories about oceanic heat waves, coral bleaching, algal blooms and fish kills, most of us are keenly aware that as the climate changes and biodiversity suffers, the web of life in the ocean is inextricably linked together, and that the health of this web is crucial to our own survival on this planet.

Simultaneously, we live in a time when the ocean itself is more connected than ever to our daily lives. The average person now eats almost twice as much seafood as people did half a century ago—and as much of that food today is farmed through aquaculture as is caught from the wild.[1] A portion of the oil that we use to power our cars and warm our homes comes from offshore drilling, summoned from deep beneath the seafloor. In the future, wind farms off the coast may help ease our transition away from this fuel, while several island nations are preparing to mine their seabeds for the valuable minerals we need to run those turbines and build electric cars, as well as our phones and computers. The list goes on and on: Our sneakers, kitchen gadgets, bedding, and children's toys often arrive at our doorsteps after crisscrossing the seas on container ships. We've discovered new treatments for cancer, malaria, HIV, and drug-resistant bacteria in deep-sea life.[2] As populations increase and water scarcity grows with climate change, coastal cities and nations are investing in desalination plants to collect drinking water from the ocean. And for many countries, ocean tourism has become a primary way of life, valued at trillions of dollars worldwide.[3]

In short, none of us can still pretend that the ocean is simply some distant, empty space that doesn't affect us. Even more than that, we *shouldn't* pretend this. The ocean can be a tremendous help to us, providing medicine, power, food, and livelihoods. Simply by existing, the ocean is moderating the effects of climate change, absorbing around 30 percent of all CO_2 released into the atmosphere;[4] model estimates suggest that without the combination of both land and ocean carbon sinks, carbon dioxide levels in our atmosphere would be nearly 50 percent higher than they already are today.[5]

But the ocean has its limits. On the carbon side, CO_2 uptake may already be slowing as ocean waters become more acidic.[6] This uptake could decline even further as the ocean heats up, since warm water stores less dissolved gas than cool water.[7] (We're already beginning to see the impact this will have, as cool-water animals shift their habitats north to escape the inexorable heat, while organisms in northern latitudes, where acidification is most severe, are stressed by increasingly acidic waters.)

We see similar limits when we consider the benefits the ocean could provide: whether a physical biochemical limit, like CO_2 absorption or the amount of fish it can support, or a limit that science suggests we should place on ourselves, to avoid destroying the very thing we're relying on.

This fact serves as a theme for the next two chapters: The ocean can be a tremendous help, a vital mitigator of humanity's biggest problems, but we need good science and careful policy to keep the ocean healthy even as we utilize it. In these chapters are three examples of ocean industries—aquaculture, ocean energy, and deep-sea mining—that could deliver needed resources to humanity. But each of these industries is at a different stage of scientific understanding, societal acceptance, and structural support in policy and law. By examining how these different factors intersect for these industries, we can better understand the sort of careful decision-making that will be needed to use the ocean while protecting the life that it holds—and how the technology that we design and rely on can help us to mediate that choice.

* * *

This thought was on my mind on a summer day in 2015, when I traveled by plane, by helicopter, and finally by boat to reach a seemingly unremarkable spot off the coast of Panama, where there is usually nothing but ocean. It took more than an hour for our destination to come into view: an enormous ring of nets with a pitched top, like a transparent circus tent, bobbing gently in the offshore swell. Pulling up alongside it, we could see the occasional flash of silver beneath the surface, the flick of a fin breaking the waves.

This ocean-going tent is known as a SeaStation, and within it, hundreds of silvery cobia—a meaty, flaky whitefish—were circling in the clear open waters of the Pacific. The fish were put here by Open Blue, a US-based aquaculture organization operating in countries around the world. Normally, cobia live on their own; you won't find them schooling in large numbers, which means that there has been little opportunity for a wild fishery. Open Blue's goal has been to raise cobia in large, easy-to-harvest numbers, to introduce this tasty fish to the seafood market, and to do so in a way with as little impact on local ecosystems as possible. This last point is why we found ourselves looking down at these fish in this place more than eight miles offshore, so far away that the coast had diminished to smudges on the horizon.

Earth's population, as ever, is growing, with no signs of slowing down, although the locations of the largest birth centers are shifting. By 2024, the global population passed 8 billion people, and is expected to surpass 9.5 billion by 2050. Meanwhile, its demographics are changing as large, formerly poor countries like India and China develop new wealth.[8] Consistently, one of the first things we see, as countries grow richer, is that their citizens introduce more meat to their diets. This fact creates a bit of an ethical quandary. Industrially raised farm animals bring with them a host of environmental problems, including but not limited to planet-warming methane emissions (from cows' own *emissions*, to put it politely), water pollution from runoff, increased possibility of antibiotic resistance, and the moral dilemma presented by raising animals in crowded, uncomfortable, often distressing spaces.

Yet trying to tell developing countries that they shouldn't eat more meat, after developed countries have enjoyed the privilege for so long, can feel hypocritical, just as it does when we say they should pull back on fossil fuel use for the cars they can now afford.

Increasingly, research is focusing on how we can produce food for this booming need and do so sustainably. A greater emphasis on local agriculture, developing new animal feeds that help cattle produce less methane, and returning to more holistic ways of raising the creatures that become our food all fit under this umbrella. So does aquaculture, the practice of growing and harvesting fish within a contained system—and it's a field with both great promise and great peril for our oceans.

This idea isn't new: There's archaeological evidence of fish ponds in what is now China (starting in 3500 BCE), Egypt (around 2000 BCE), and Italy (likely starting in the first century BCE). As far back as a thousand years ago, Indigenous Hawaiians developed advanced, aquaponic-like systems that grew fish alongside seaweed or vegetables like taro;[9] and at least thirty-five hundred years ago, First Nations tribes in the Pacific Northwest constructed "clam gardens" that created the ideal conditions for the mollusks they harvested to feed their people.[10] Yet aquaculture on the scale we're seeing today, and the scale it may need to be to feed the world, is on a completely different level. In 2020, the Food and Agriculture Organization (FAO) of the United Nations estimated that aquaculture facilities worldwide produced almost 181 billion pounds (82.1 million metric tons) of fish, a roughly five-fold jump in the last thirty years—and one that brings the proportion of farmed fish and wild fish consumed by humanity to nearly 50/50.[11]

In many ways, the aquaculture revolution heralds a healing progress for our oceans. Globally, around one-third of fish stocks are being eaten faster than reproduction can replace them. In some places the statistics are even worse: In the North Atlantic, more than 40 percent of fish are classified as over-fished, meaning they can't reproduce fast enough to replace what's being caught. In the Mediterranean, around 90 percent of species are overfished.[12] The process of fishing can be damaging, too: Bottom trawling, in which weighed nets are dragged across the seafloor to catch shrimp, prawns, and bottom-dwelling fish like flounder and cod, destroys habitats by ripping up plants and destroying burrows, corals, and sponges—and can leave nets behind that entangle animals.[13] It also kicks up sediment that shuts out sunlight needed by marine plants, and disrupts another way the ocean helps mitigate climate change: According to one study, bottom trawling annually re-suspends substantial amounts of carbon, although the exact amounts are in dispute.[14] Additionally, while many governments have become more stringent in passing regulations that prevent bycatch—when animals not meant for the dinner plate, such as sharks, birds, sea turtles, and dolphins, end up in fishing nets—it remains a problem, especially in countries with less rigorous fishery oversight.[15]

With all of this to consider, perhaps it would be better to simply raise fish the way we grow lettuce, chickens, or apples, removing pressure off of their wild cousins. After all, evidence shows that if fishing is carefully managed, wild species can re-cover, sometimes even rapidly.[16]

If only it were that simple. Aquaculture in coastal waters has become a lightning rod of criticism, particularly when used to raise finfish like salmon. Non-native fish can and have escaped from these pens, creating competition with local species; even when native species raised for consumption escape, the reduced genetic diversity in these escapees may make local fish less fit through interbreeding. Additionally, crowded pens can spread diseases and parasites, both within the farm-raised stock and out to wild fish. Just look at the salmon aquaculture industry, which has spent years and millions of dollars figuring out how to treat sea lice, a crustacean parasite that at best creates lesions that make the fish unmarketable, and at worst can kill an entire stock.

Historically, the primary treatment for sea lice was a chemical cleanse that risked killing the fish themselves and meant de-laying a fish's harvest to avoid passing that chemical on to the consumer. In some populations, sea lice started to evolve a resistance to these chemicals.[17] In Norway, home of the most technologically advanced aquaculture industry in the world, a multi-tier solution is emerging. First, better understanding of the coastal environment allows management of farm densities in the fjords to minimize transmission of sea lice between adjacent facilities. Next, using bubble circulators to bring colder water to the surface in and near cages allows farmers to make the waters less hospitable to sea lice, which like the bright, warm, highly oxygenated surface waters. Finally, the ability to completely submerge cages offers the prospect of removing fish from the surface habitat of the sea lice and provides other advantages such as protection from rough surface conditions during storms. But all of this comes with significant technological complexity. With aquaculture's growth, it's easy to forget how young an industry its modern incarnation is, compared to ten millennia of farming. The cycle of experimentation, learning, and improvement is in its early days.

Then there's the problem of harmful algal blooms. Fish waste, uneaten food, and dead fish from these pens can accumulate nutrients like nitrogen and phosphorus in the water. These nutrients are the same ones that enable plants to grow, and in the ocean, they do exactly that: provide food for algae to grow. Too much of these nutrients can cause algal blooms so large that they deplete oxygen in the water or feed species that create toxins.

Harmful algal blooms have become the bane of many fish farms. Fishes' natural defense when faced with low-oxygen water or toxins would be to swim away, but in a contained pen, that's not possible. Often, harmful algal blooms can kill most or all of a fish farm's stock, leading to losses in the millions of dollars. Larger events can also have knock-on effects, like the devastating pair of blooms that swept through southern Chile in 2016, killing millions of farmed salmon and making wild shellfish and farmed mussels inedible.

However, saying that the farms *caused* these harmful algal blooms is tricky business. There's scientific consensus that warming ocean waters are making harmful algal blooms more likely. Yet because aquaculture companies often don't gather information on what local waters looked like before they install fish pens, there isn't a baseline that can demonstrate whether the excess nutrients around fish farms are the exact cause. Would the bloom have happened if the farm wasn't there, and the stationary pens simply made an otherwise unseen bloom visible? Or did the farms provide the other ingredient needed for algae to take advantage of warming waters?

These problems can be worse in certain regions, such as in countries with fewer environmental regulations, which can allow aquaculture farms to release pollutants and nutrients with little oversight. And certain species are known as being worse than others. Shrimp farming in Asia provides an example of both: Countries like Indonesia, India, Thailand, and Vietnam have historically overlooked the destruction of mangrove forests to create inshore pens, which then concentrate the nutrients from feed and chemicals added to prevent disease. When the shrimp are removed, all of that polluted water is freely released into rivers and coastal habitats. That shrimp goes on to feed the world: Asia accounts for 85 percent of global shrimp aquaculture production, and in the United States, more than 90 percent of the shrimp we eat comes from outside of the country.

As is probably clear by now, aquaculture's problems are complicated, and broader societal changes, like stricter regulations, are an essential first step in many places. Yet many of the problems with aquaculture come from the fact that they are near the coast. These shallow, less-circulated waters concentrate any pollutants and put fish farms in close proximity to human activity, vulnerable ecosystems like reefs and marshes, and other populations of fish that escapees can contaminate or breed with. Simply changing *where* we farm these fish could, perhaps, tackle multiple problems with one solution.

One option is moving aquaculture onshore, where it theoretically has no way of polluting local waters, and where land is the only limitation to the size of the farm. Yet the waste problem doesn't go away because the farm is on land. The other is moving aquaculture *way* offshore, far from the more sensitive and highly populated ecosystems on the coast. There, stronger ocean currents can sweep away and dilute any human inputs as well as, some research suggests, lead to fewer parasites. But solutions to complicated problems are rarely simple.

"I think most people in our field would probably agree there is no one-size-fits-all for aquaculture," says Halley Froehlich, a researcher studying the sustainability of aquaculture and its future under climate change at the University of California–Santa Barbara. "And that is really reflective of how diverse aquaculture is actually, in the real world: everything from if you have sticks and line you can grow seaweed to something as complex as the huge offshore structures used to raise fish in Norway and China."

Froehlich says that the right choice of aquaculture for a location will likely depend on many factors, including the ocean conditions in the region, the level of technology available, how much fish is being grown, and the species involved. For instance, raising seaweed or shellfish like oysters in coastal waters doesn't pose as many problems as raising finfish. These species need relatively little human intervention and can actually help clean pollution and excess nutrients from the water; a single adult oyster can filter up to 50 gallons of water per day, removing excess nutrients and pollutants alike.

Each type of aquaculture will have to consider both the benefits and the tradeoffs. Froehlich's research has shown that offshore aquaculture reduces the risk of disease in fish and improves their living conditions, but escapees still remain an issue. While theory suggests that it would be less likely for these runaways to reach and interact with coastal populations, there's little research on the topic so far.

On land, contained aquaculture operations tend to boast much higher yields than farms in the ocean, but they're also much more expensive, and require a lot of energy. Onshore aquaculture also has a higher tendency toward catastrophic failures; a single technical problem or a case of contamination can wipe out an entire generation of fish. Finally, Froehlich points out that land-based aquaculture can continue to threaten biodiversity, if open land is converted into these facilities.

* * *

My own interest in aquaculture came from my growing interest in the ocean as a habitat. Was aquaculture a viable solution to overfishing? Could offshore aquaculture mitigate the environmental impacts of intensive near-shore aquaculture? And if so, what would it take to make it economically competitive? Despite the benefits of offshore aquaculture, the state of technology today still requires human caretakers to take frequent boat trips offshore to feed the school, monitor its health, remove any dead fish, and clean off layers of algae that can grow on the pen—and until electric-powered boat engines are more common, each of these trips releases emissions that contribute to climate change. I was particularly interested in how sensors and scientific understanding of fish farming could take offshore aquaculture from trial and error to a scientifically managed enterprise.

This is what brought me to the coast of Panama. Open Blue's cobia pen was designed by Innovasea, a company that creates technology for offshore aquaculture; their technology was compelling enough that I became one of their board members. Innovasea has invested in technology that allows them to remotely raise and lower their pens in the water column, allowing them to move fish to safety if rough weather is coming. Additionally, many of their pens are monitored by cameras and computers that can quantify the overall population of their school and their sizes, and monitor behavior to make sure fish are getting enough to eat.

But professionals in the field eventually see it going even further. Already, researchers around the world—including in Germany,[18] Greece,[19] Canada,[20] Norway,[21] and New Zealand[22]—have created and tested prototypes of both ROVs and AUVs equipped with software to monitor the water quality around pens, keep an eye on population numbers, and examine the fishes' skin and fin health, as well as look out for fish behavior that might indicate distress or sickness. This would allow for early detection of diseases, allowing companies to get ahead of any conditions that could spread to wild populations as well as prevent the overuse of antibiotics.[23] Autonomous "crawlers" already exist that can clear biofouling organisms off the outside of aquaculture nets. Experts envision that it wouldn't be too much of a jump to give these machines the ability to recognize and repair breaks in the net, helping to cut down on escapees.

The same sort of machine vision that can monitor feeding behavior could also route back into automatic feeding systems that help limit overfeeding, preventing excess food from falling to the seafloor and becoming fodder for harmful algal blooms. These systems would also make it so that humans only had to visit the pens every few weeks, reducing the greenhouse gas emissions from boats coming and going. US government organizations, including NOAA, the Department of Commerce, and especially the National Sea Grant Program, have all shown great interest in investing in this sort of research, and it's a focus of my old lab, the MIT Sea Grant AUV Lab.[24]

None of this interest is surprising to David Kelly, formerly of Bluefin, who is today Innovasea's CEO. Smarter aquaculture is not just better for the planet, he says, but better for companies' bottom lines.

"The economics for the farmer are aligned with minimizing use of resources, minimizing waste," Kelly explains. "The largest variable cost is feed, so it behooves you to feed the fish as efficiently as you can. And the fish are sensitive to their environment, so it doesn't do you any good to raise fish in a poor environment." In other words, he adds: "The economic drivers are aligned with fish health and stewardship of resources, and of the planet."

The biggest hurdle that aquaculture faces, from experts' point of view, is buy-in from the broader public. The early image of coastal aquaculture as a dirty business that produces sick, unhappy fish is one that has stuck around and spread to the industry as a whole in many developed nations. However, this sentiment isn't universal. In a study published in 2017,[25] Froehlich examined newspaper headlines and public comments and found that the coverage in the developed world, and particularly the United States, tended to frame aquaculture more negatively than in the developing world.

To Kelly, this reflects two facets of reality that he's observed in the industry. He feels both that views of aquaculture's sustainability trail its reality by at least two or three decades, and that if developing countries don't catch up, they'll end up chasing the rest of the world. "This is a regulatory problem and a social license problem, not a technology problem," he says. It's also a funding problem for developing nations, where little cash leads farmers to do things the easiest way, which isn't often the most sustainable way.

Froehlich's 2017 study found that positive coverage of aquaculture in the media *has* been increasing over the last twenty years. In order to continue that trend, Froehlich says, regulators, NGOs, and aquaculture organizations should build on the warming public attitudes toward the industry by bringing stakeholders—coastal residents, fishermen, local governments, and other area industry players—into the conversation. "From my perspective, the context matters a lot: Who is at the table deciding, and who has a voice on what is being done for what reason, and who benefits, really matters and should matter," she says.

At the same time, these experts emphasized that part of the struggle will be to show the public that even if aquaculture does have environmental impacts, they need to be weighed against the fact that they are still an improvement on the existing seafood system—the lesser of two imperfect approaches.

"There's this interesting tension point on how to reconcile these voices of concern and validate that feeling, but also be cognizant you don't get lunch for free," Froehlich adds. "There's always going to be an impact. And seafood should be more thought of in the collective of our food system, and not just on the seafood merits alone."

The fish farm that I saw in Panama, and others like it, serve as a potential model for how all of this might work. The organization iAlumbra—a collective of nonprofit and for-profit aquaculture companies, including Open Blue—has made it its mission to invest in offshore aquaculture specifically in the developed world and in food-insecure communities, in an effort to show that these farms can be developed sustainably and in cooperation with local communities. The vast majority of the employees of these aquaculture farms come from and live in the communities where they are based. And these employees have a real role in the direction that iAlumbra's farms take, from the fish that they grow to the projects they invest in.

"The business models we've seen, especially for the last thirty years around maximizing profits for shareholders, is not sustainable," says Jamie Cherney, president of iAlumbra. "It doesn't take into consideration the other constituents, including the planet and employees. We really are trying to approach business differently in how companies deploy resources, in a way that respects the community."

In Kona, Hawaii, 90 percent of food consumed comes from off-island. The community there saw tourism-based incomes shatter, and the supply chain for much of their food drop off, during the COVID-19 pandemic. But this isn't how it has to be.

iAlumbra has a farm in Kona growing kampachi, a native Pacific fish popular in the sushi market; in addition to providing income to Kona-based workers, half of the fish produced there deliberately stays on-island for local consumption. In Panama, iAlumbra has worked with locals to dig wells for much-needed clean water. Across several of its farms, it's working with the Nature Conservancy to bring oyster farms to accompany fish farms, which will provide further income as well as clean local waters and store carbon. And in Baja California, Mexico, the company not only grows totoaba for consumption and sale but also participates in releases to help repopulate the native population of this fish in the Sea of Cortez, where its population was devastated due to the value of their swim bladders in Asian medicine.

Cherney sees this model as a way to demonstrate that aquaculture can turn around its negative reputation, in places where it's only known for polluting coastal waters in pursuit of profits. It's also a way to show countries that are behind on adopting aquaculture—like the United States—what's possible, even when there is pushback against the field about its environmental bona fides. "It's kind of like: 'or what?'" she says. "I absolutely understand not wanting to have environmental harm, but if we're not setting the standards ourselves, then we're probably letting other countries set the standards, and they might not be as responsible or planet-oriented as we want."

These conversations are not only limited to the food systems that come from our oceans. In fact, perhaps nowhere is the conversation more heated right now than in the energy space. As society looks toward a future focused on releasing fewer carbon emissions, the ocean again stands to provide us with new avenues for how we power our world. But how we obtain this power is still under consideration, and the methods by which we do so could fundamentally change the future of our oceans and their inhabitants.

CHAPTER 6

Powered by the Ocean

ON APRIL 10, 2010, Marcia McNutt—at the time, the director of the US Geological Survey—received a call from Ken Salazar, the secretary of the interior, urgently asking her to come to a meeting. An oil drilling platform had exploded into flames in the Gulf of Mexico, and the federal government was just beginning to wrap its collective head around what would come next for worker safety, for local shorelines, wildlife, and fisheries, and how to stop the flow of oil that was now gushing into the sea.

Shortly after the meeting ended, Salazar asked McNutt to go down to the Gulf for a survey of the site, scheduled to take about three days. She stayed for four months. The work the Geological Survey and other federal agencies did after Deepwater Horizon would permanently change the way the federal government responded and conducted science following disasters. For many of us in the marine community, the work that went into investigating, plugging, and cleaning up from the spill would be yet another wake-up call when it came to the link between our oceans and our energy future.

Mere hours after British Petroleum (BP)'s Deepwater Horizon well blew, McNutt put out a call to the oceanographic community for help. Oil was now gushing into the Gulf of Mexico more than a mile down, deeper than any other blowout seen before. Yet she had been told that the National Oceanic and Atmospheric Administration (NOAA) did not have the survey platforms needed to understand the extent of the spill under the sea surface. What kind of tools did we have, she asked us, to understand what was happening?

The platforms Marcia was looking for had been funded by ONR and were found in a few research labs around the country—a legacy of AOSN. Almost immediately, multiple teams equipped with marine robotics were on the scene to help. MBARI brought the newly designed *Dorado* AUV, developed by my lab and the descendant of a vehicle I had brought there from MIT. *Dorado* was equipped with artificial intelligence that allowed it to decide where to sample next based on the water samples it "gulped" in, allowing it essentially to follow the oil. The algorithm was developed by Yanwu Zhang, a member of my MBARI lab, to sample phytoplankton patches. Zhang accompanied the vehicle to the Gulf, knowing full well that his software would need to be tuned to adapt to its new mission of bringing back oil samples from the deep. The *Dorado* map was the first to show that the plume of oily water from the break wasn't all rising to the surface but, instead, gathering on water layers with different density. "The amount of oil at the surface was but a mere fraction of the oil coming out," McNutt explains.

Soon after, two researchers from WHOI, Rich Camille and Chris Reddy, had the idea to put an acoustic Doppler current profiler—which measures how fast water is moving throughout a water column—and a mapping sonar onto one of the ROVs being used to look at the broken wellhead. The combined data from those instruments allowed the Geological Survey to put the first accurate number on the flow rate out of the well, which also told them how much oil had already been spilled. That data, McNutt said, triggered the decision that drilling relief wells to reroute the flow of oil wouldn't be enough; the wellhead needed to be shut immediately. "We couldn't wait for the relief wells, which were still weeks away," she says.

Many attempts sought to close the well in the weeks that followed. These included using golf balls and rubber balls to plug the hole, and a near-disastrous attempt to use a methane-filled steel balloon to close the well. This attempt nearly ended in another explosion when methane ice made the balloon buoyant and sent it rocketing toward rescue boats on the surface. But on July 15, just before 2:30 p.m. local time, BP sealed the valves on the 'capping stack' structure, which had been attached two days earlier to the damaged well: For the first time in 87 days, oil stopped flowing into the Gulf of Mexico.[1] Over the next two months, relief wells pumped cement into the flow channel and the base of the original well, sealing it permanently.

After Deepwater Horizon, many of us working in the marine science field—indeed, many people in the public who followed the events—thought that this catastrophe would change how our country manages the ocean energy industry. After all, eleven workers had died in the oil platform's explosion, which injured seventeen others; more than three million barrels of oil were released into the Gulf, the biggest oil spill ever in US waters; and fisheries along the Gulf Coast closed for more than two months, losing the commercial fishing industry between $94.7 million and $1.6 billion and as many as 9,300 jobs.[2] The wildlife toll in the region was immense, and it's still being accounted for. The Audubon Society estimates that 1 million birds were killed,[3] while NOAA documented huge mortalities among turtles and dolphins, including the immediate deaths of more than 20 percent of all juvenile Kemp's Ridley turtles and a 50 percent decline in the population of bottlenose dolphins following the spill.[4] And thanks to surveys by ROVs and AUVs, we know the oil wreaked havoc on deep-sea life, killing zooplankton, shrimp, crabs, and mollusks, and damaging deep-sea corals that can take hundreds of years to grow.[5]

To this day, surveys around the wellhead show that it's thronged with zombie-like shrimp and crabs covered with tumors and parasites. The theory is that decomposing hydrocarbons smell, to them, like a mating signal, which lures these creatures into a sticky trap that then prevents them from molting and keeps them there, waiting for a mate that will never come and slowly falling apart.[6] The cause of the spill was chalked up to shortcuts and last-minute changes taken by BP to keep its budget down, which led to the failure of a cement cap containing oil with the well.[7] In 2020, *The New York Times* conducted interviews with every member of the bipartisan committee convened by the federal government to investigate the cause of the Deepwater Horizon disaster. All seven members said that many of their recommendations after the disaster, such as developing a drilling safety agency and passing legislation to protect whistleblowers reporting violations aboard oil rigs, were never taken seriously—likely because they were costly and would have required significant political will to pass new laws. These experts said that another catastrophic spill was not only possible, but that the United States was only "marginally better prepared" than it was in 2010.[8]

It's not all bad: On the commercial side, oil well–containment capabilities have improved significantly. The Department of the Interior also restructured the Minerals Management Service, separating the enforcement and revenue-generating functions into different offices, removing an inherent conflict of interest in regulating energy companies. However, the US Congress has enacted almost none of the drilling safety recommendations that followed Deepwater Horizon. Some of the safety rules established after the spill were later rolled back in response to industry concerns about cost and feasibility. The number of drilling rigs in American waters is declining,[9] but oil prospecting has continued—and shifting political currents could see more drilling open in the near future.[10]

One big lesson for me was the unsettled nature of the science of oil spill response. Simple questions, like "how much oil is being released?" turn out to be very hard to answer. Further, critical decisions, such as whether to inject dispersants into the oil streaming out of the seafloor, had to be made despite significant scientific ambiguity. As I learned years later at an oil spill conference we hosted when I was at WHOI, the experts are perfectly capable of arguing either side of the issue. Of course, responders don't have the luxury of waiting for a methodical and careful scientific study—they have to work with what imperfect knowledge they have, which is why it's so important to do the science before there is a crisis. One of the more consequential outcomes of the disaster was the creation of the Gulf of Mexico Research Initiative (GoMRI) which was established to study the impacts of oil spills on the environment and on public health, and to advance the science and technology for oil spill response.[11] This was established by BP with a commitment of $500 million over ten years. It funded more than a thousand scientists—and their students and staff—on projects giving us our clearest view of the impacts both of the oil spill itself and of the efficacy of response efforts.

The risks of offshore oil drilling to the ocean are, of course, not limited to American waters; it's a huge global industry, from the Santos Basin off of Brazil, to the enormous wealth of the North Sea, to the oil claims that have led China to exert their military might over the South China Sea. However, where the Deepwater Horizon disaster is particularly instructive is in demonstrating what it was *not*. The explosion at BP's platform occurred around 25 miles (41 kilometers) off the coast of southeast Louisiana, in the center of one of the most industrially active marine regions in the world. It was surrounded by ports, including New Orleans, Biloxi, Gulfport, Pensacola, and Panama City, all with infrastructure capable of supporting the ships, helicopters, planes, underwater vehicles, and people needed to respond. Catastrophic as this spill was, you could almost call its location convenient.

The same will not be true of any offshore drilling accidents that happen in the Arctic, an oil source to which the world is turning its eyes, ironically, as climate change shrinks its annual ice cover. While Norway, Iceland, Russia, and Greenland all have deepwater ports in the region between the Atlantic and the Arctic Oceans, these are much fewer—and farther apart— than one finds in temperate seas. The Russian Arctic has a series of developed ports well supported by the country's icebreaker fleet, but many of these ports require a long river transit for other vessels to reach them, and ships often must be accompanied by icebreakers to transit waters that are frozen much of the year. Near the Bering Strait, along the North Slope of Alaska, and in northern Canadian waters, there is almost no support infrastructure.[12]

This problem is of particular interest to me as an AUV developer, and as someone who has tailored my relationships with the oil and gas industry around ensuring the safety of subsea systems and developing emergency response capabilities. These relationships started early in my career: The business we built at Bluefin counted commercial offshore survey companies—which carry out surveys for oil and gas infrastructure—as big clients, although the US Navy was our largest by far.

Yet the enormous challenges associated with understanding what was going on below the surface in the Gulf of Mexico led me to ask: How will we even know what is going on if there is an accident in the Arctic? This led me to start a Department of Homeland Security–funded initiative, which kicked off in 2013, to adapt the *Tethys* vehicle that we designed at MBARI as a rapid-response vehicle in the event of an Arctic oil spill.[13]

Congress might have failed to follow through on measures to prevent the next Deepwater Horizon, but those of us working in subsea technology could begin to organize around twin challenges: preventing future incidents, and providing a much more effective response should a disaster occur. In the years following Deepwater Horizon, I continued to shift how I worked with this industry. This change accelerated with my move to WHOI, where there was work being done on technology that could help the industry with maintenance and inspection of its equipment and its underwater wells.

One of the biggest challenges here stems from offshore oil's success. Deep-sea wells are so productive and long-lived that they can continue to produce abundant oil even as their subsea infrastructure approaches the end of its life. An aging well should therefore have more frequent inspections and more maintenance. But performing these activities with a deepwater ROV is very expensive—and we suspect from the Deepwater Horizon experience that there are companies that will cut corners. The concern is that there may be another disaster waiting to happen.

The solution is to cut the tether. The high cost of ROVs is largely driven by the cable connecting it to the surface, as tethered operations in deep water require ships that can hold an accurate position in the unpredictable conditions of the open ocean. In other words, an expensive ship. However, an untethered vehicle can be operated and maintained from a much less expensive ship, and maybe from no ship at all. Further, without a tether one can operate in much more challenging weather conditions. In my time at sea, I have frequently operated AUVs from fishing boats, which are plentiful and economical. But to work in an oil field we need a new type of AUV, one that can operate close to the extremely expensive equipment on the seafloor with high precision and safety. Ultimately, it requires autonomous manipulation—a particularly exciting area of development that continues today.

My connection with the oil and gas industry over the last decade has focused on creating vehicles and other technologies that make inspection and maintenance affordable and easy, so that these wells can operate safely and cleanly for the rest of their lives. There is increasing demand for these vehicles, though at present they are concentrated in a few hotspots worldwide.[14] Given the growing interest in fossil fuels below the Arctic, investment in more autonomous monitoring and response would be one way to mitigate the damage that could be caused by a spill in inaccessible northern waters. We will surely need this sort of technology as deepwater infrastructure ages.

Just as advances in ocean science and technology funded by the Navy have had broad impacts on the ocean science enterprise, we can see that the technologies being developed to build, sustain, and monitor oil activity in the deep ocean can set us up for a cleaner future. Many of the same technologies that are useful to offshore oil wells can be used to maintain wind farms planned for coastal waters around the world. In fact, wind farms are a simpler problem in many respects; they operate in shallower water depths, and the complicated parts are all above water. With growing momentum for a global shift toward green energy, the same technologies that kept fossil fuels flowing will maintain their replacements when these wells close for good.

* * *

Most climate experts agree that in order to prevent the worst consequences of climate change, we need to limit the planet's warming to within 1.5 to 2 degrees Celsius (2.7 to 3.6 degrees Fahrenheit) above preindustrial levels.[15] To do that, there's no question that humanity needs to shift our energy system away from releasing so many greenhouse gases. But there's no right answer when it comes to *how* we make that shift.

At present, it's still unclear how the proportion of renewables like wind and solar, nuclear energy, carbon capture, and other still-unknown technologies will settle out, and analyzing those potential futures is well beyond the scope of this book. What we can examine, however, is the potential contribution that ocean energy will make to this cleaner future—and how the same sorts of technology we've discussed so far can not only help implement these new forms of energy but also ensure that we don't repeat the mistakes of the past, disregarding ocean ecosystems in favor of faster, cheaper, and easier-to-procure energy.

"I would suggest that the ocean is going through a marine industrial revolution—it's happening before our eyes right now," says Andrew Lipsky of NOAA's Northeast Fisheries Science Center, who co-chairs the International Council on the Exploration of the Sea's Fisheries and Offshore Wind Working Group and has been working in fisheries and renewable energy for his entire career. Over the course of this career, he says, "you learn how we have great examples of how we haven't fully considered the scientific needs to account for the costs of our actions, and effectively monitor the interactions—particularly when we discover a new natural resource and prepare to exploit it for human use."

Offshore wind is an industry that's seeing exactly this sort of boom: Experts say that wind development, and interest, is growing faster than science can keep up with it. There is a lot of potential but also much that we don't know about the impacts of wind farms in the ocean. There's little question that these farms are a net positive for the climate—but by understanding how they impact the oceans around them, we can take steps to make sure they're a net positive for everything that shares those waters.

Some of the biggest lingering concerns are related to underwater sound pollution, created both by the construction of these farms and by the constant vibration of the structures from their spinning blades. Scientists still aren't sure how this could affect animals like whales, which use sound to communicate, and the movements of migratory animals like fish and turtles. Turbines also pose the potential for injuries or fatalities to seabirds. On the policy side, too, state and federal agencies are largely unprepared to deal with fishing communities that will suddenly find their historic fishing grounds inaccessible; industry and government will need to set up legal structures set up to compensate those who've been displaced or devise a new permitting scheme to allow them to fish elsewhere. Only two offshore wind farms currently operate in US waters, and they're small enough that fishing disruptions have been minimal. This would not be the case with farms of the size needed to provide significant energy generation.

This research is not only essential for the scientific continuity of fishing regulations, but is also more important than ever when a wind farm is put in. For example, research on the Block Island Wind Farm, which has been in place since 2015, showed that the bases of the wind towers quickly became colonized by mussels and attracted shelter-favoring fish like black sea bass—in effect, creating an artificial reef that could help bolster fisheries in the area, a benefit that has so far been corroborated by research on other farms. Yet one lingering question around these farms is whether they actually enable reproduction and create more fish (by providing them with more habitat than they had before), or just encourage fish to move from one place to another, with no net increase in the overall population. This uncertainty makes it hard to say that putting a wind farm in one place won't impact valuable fisheries in another place.

Could technology help address some of these lingering concerns? Unsurprisingly, underwater vehicles are part of the suite of tools being used to find out. WHOI has been working with NOAA to develop an AUV using cameras and, in the future, machine learning to carry out fish surveys, with the computer automatically identifying the fish species and other underwater features that researchers today annotate by hand. [16] They hope to integrate other remote-monitoring sensors into wind farms, like eDNA systems, that would allow even more detailed views of how the new installations alter local ecosystems. More thorough surveys of wind farms could help mitigate the lack of long-term data that plagues this industry; even in Europe, where offshore farms have been in place for decades, there is scant work on how these structures can impact marine communities over the long term.

There has also been a push within the wind industry to change the way wind farms are built, moving away from wind towers and instead toward floating, moored turbines. Fixed turbines can't be placed in water deeper than 180 feet (55 meters), so this change could allow wind farms to move into much deeper waters—out of the way of productive coastal ecosystems, and where the lack of a solid base would create less interference and less sound pollution in the water column. These farms would require much longer underwater cables, which come with ecosystem impacts of their own that will also need to be studied and monitored.

The ocean energy space is replete with new ideas and new technology, including a growing body of research on tidal energy and wave energy, still considered the "Wild West" of the ocean power field. In many ways, this field is reminiscent of the early days of AUV development; the rules had to be written as we went along. In the case of ocean energy, however, the stakes are much higher, and the consequences for the environment and local communities greater if we don't consider them ahead of time.

Yet I would argue that there is no better time to take on those stakes. Manufacturing is undergoing a sea change as investors demand more transparency and action around the climate vulnerability of different industries; in the United States, the Securities and Exchange Commission adopted new rules in 2024 to require climate-related disclosures—which positions clean energy to stride ahead.[17] But these industries are financially vulnerable, and the bottom line still rules. The number one way to kill investment will be to create uncertainty around whether new installations will ever be approved, in response to environmental concerns or conflicts with the public.

To prevent this, we need well-funded marine science that can leverage and motivate the technology, using it to observe the impacts and develop the scientific understanding needed to prioritize risks. We need a regulatory process that adapts to our changing understanding, while providing permitting pathways that minimize financial uncertainty for investors. And we need the growing community of marine tech entrepreneurs focused on studying the impacts, not just the economics, of installing and operating new energy farms.

Michael Lawson, the group manager of the water power R&D group at the National Renewable Energy Laboratory, is one of the people trying to strike this delicate balance in the field of wave energy. The mistakes of the past are indeed part of how people in his industry plan for the future—and avoid overpromising for new technology. "It might seem like charging ahead is the fastest way to make progress, but we've learned that's actually not true. One misstep can set you back years," he says. "People are very cognizant of the fact that if you charge ahead too fast, you can give your technology and industry a black mark that could set them back five to ten years—or worst case, be a death sentence."

THE PROMISE AND PERIL OF DEEP-SEA MINING

Similar conversations are happening today around another resource: the metals that we need to power our phones, electric cars, and indeed wind turbines themselves. Some of these valuable minerals are quite literally scattered across the deep seafloor, creating a tantalizing prospect for mining companies and the countries that might benefit from them. But accessing those minerals is a complicated proposition.

For more than 150 years, oceanographers have known that the deep ocean was home to rocks called polymetallic nodules: unremarkable-looking rocks on the outside that hold valuable metals within. The *Challenger* expedition, in the 1870s, pulled up some of the first examples from the North Atlantic. Over time, scientists realized that these nodules were relatively abundant, scattered across the flat, empty, perpetually dark seafloor between one and four miles deep known as the abyssal plain. This ecosystem is estimated to cover an astonishing 60 percent of Earth's surface and 83 percent of the total ocean seafloor.

Over the span of centuries, minerals slowly precipitate out of seawater itself, and as they fall to the abyssal plain, slowly accrete around rocks or shells to form these nodules. Many of those accreting minerals, like manganese, nickel, copper, and cobalt, are highly valuable, but it's always been far too expensive and difficult to reach them.

Fast-forward to the present: From a technical standpoint, many of the challenges of working on the abyssal plain have been solved. Companies interested in deep-sea prospecting have been conducting tests that appear to place them just a few years away from having the technical capability to do so, using surface-controlled ROVs that work essentially like deep-sea vacuums to suck up the nodules from the seafloor. The nickel, cobalt, and manganese found within these nodules are a valuable resource in the face of climate change, as they could be used in batteries for electric cars, solar panels, and wind turbines. But while we have the technology we need to get at these resources, we don't yet have the regulations or societal agreement needed to safely do so, nor the science to say definitively what the domino effects might be on the oceans and marine life.

The primary legal structure around ocean uses comes from the United Nations Convention on the Law of the Sea (UNCLOS), a 1982 agreement intended to govern all activities in the oceans. One of the most significant provisions of this agreement was formally establishing what parts of the ocean a country could claim as its own and where only that country had rights over natural resources—an area called its exclusive economic zone, or EEZ. A country's EEZ extends 200 nautical miles from the low-water mark of its shoreline. Everything outside of that border is considered the high seas, the two-thirds of the world's oceans that lie outside the national borders of any country.

The 1982 agreement included general language about safeguarding resources from the high seas, but it didn't include any official protections for marine life beyond national borders. Nonetheless, it did address one specific aspect of exploiting the high seas: the ability to extract valuable minerals from this vast expanse. Part XI of UNCLOS established the International Seabed Authority (ISA) to approve and manage mining in the high seas and responsibly distribute any profit from those actions.

Notably, one major nation did not agree with the Law of the Sea. The United States, under President Ronald Reagan, refused to sign on to UNCLOS because of the possibility that Part XI might limit the ability of private companies to mine the deep sea. Although observed by the United States, UNCLOS remains unratified to this day, with opposition motivated by diverse concerns including potential loss of sovereignty, suspicion of the United Nations (which is the administrating body), and potential revenue- and technology-sharing requirements.

In the three decades since its establishment, the ISA has still not issued official regulations governing seabed mining. In fact, the ISA has been debating the mining code, the regulations that will govern the first permits for deep-sea mining in in-

A New Agreement to Protect the High Seas

On Saturday, March 4, 2023, Singaporean diplomat Rena Lee stood at the United Nations headquarters in New York, in front of negotiators from nearly two hundred countries, and announced to everyone gathered: "The ship has reached the shore." After two decades of diplomatic work, two weeks of discussion, and in the end, more than thirty-six sleepless hours of final negotiation, the people in that room had agreed on the text of the first agreement to protect marine life in the high seas—and help to fairly distribute its resources.

The new High Seas Treaty, formally adopted by the United Nations on June 19, 2023, was an attempt to address some of the gaps in existing international law around the oceans—namely, the United Nations Convention on the Law of the Sea (UNCLOS), which included no specific protections of marine life in areas of the ocean outside of national jurisdiction. The high seas account for 95 percent of the world's total habitat by volume, but the Marine Conservation Institute estimates under 3 percent of it is protected.

Also known as the Agreement on Biodiversity Beyond National Jurisdiction, or BBNJ, one of the new treaty's major achievements is creating a process for the establishment of marine protected areas (MPAs) in the high seas. In these MPAs, regulations will protect ecosystem services, conserve valuable cultural heritage, or pursue a conservation goal. Importantly, the MPA creation process does not require consensus by all parties; requirements for consensus already delayed the creation of MPAs around Antarctica, owing to vetoes from China and Russia. Observers hope that this facet of the treaty may help the United Nations achieve the goals set out by the 2022 Kunming-Montreal Global Biodiversity Framework, which aims to protect at least 30 percent of Earth's land and water by 2030.

The BBNJ is one of the first and few international agreements to put equity at its center. The agreement lays out provisions to ensure that all peoples benefit from the high seas, which are recognized as the "common heritage of humankind." One of the major points negotiated in the treaty is the process for how the high seas' marine genetic and biological resources—which might be used to create everything from new medicines to cosmetics—would be used, including a monitoring and assessment process to theoretically prevent exploitation. These provisions also established a joint trust fund that would distribute a share of any profits from these products to all nations. In seeking "to achieve universal participation" in the management of the high seas, the treaty also contains language ensuring that Indigenous voices and small island communities are included. Additionally, the BBNJ establishes a mechanism that would provide technology transfer, scientific training, and capacity building to lower-income nations, helping to build an on-ramp to the high seas for countries without a well-developed marine technology sector.

Yet questions remain about how the treaty will work in practice; there are diverse perspectives. Nations still need to identify and organize the funding channels for the trust fund to share profits.

Indigenous and environmental scholars have pointed out that the treaty remains anthropocentric rather than ecocentric, focusing on humanity's right to exploit nature rather than centering the idea that nature has rights of its own. And fishing remains a thorn in the BBNJ's conservation credibility. The text has few mentions of the fishing industry, despite it currently being the largest form of industry on the high seas, and it establishes no methods by which fishing can be prohibited in high-seas MPAs. If a nation started the application process for an MPA under the current text, it would need to convince any regional fisheries management organizations that fishing should be prohibited, which would be unlikely to succeed.

Overall, much of the treaty's language is extremely broad, with few definitive outlines for how its requirements will be implemented or enforced. Marine researcher Jeffrey Marlow, a scientific advisor for the negotiations, called this vagueness "constructive ambiguity" in an article for *The New Yorker*, noting that it "is instead meant to serve as a scaffold for future initiatives."

After its adoption in June 2023, the text will need to be ratified by each of the United Nations member states before it comes into force, a process that can take many years; UNCLOS itself took twelve years to be ratified. But hopes are high that it won't take so long. Other environmental regulations have moved through the process more quickly—formulating the Paris Agreements took less than a year—and participants and observers are hoping for the BBNJ to be legally adopted at the third UN Oceans Conference in Nice, France, in June 2025.

FURTHER READING

Deasy, Kristin. "What We Know About the New High Seas Treaty." *npj Ocean Sustainability* 2 (June 30, 2023), article 7. https://doi.org/10.1038/s44183-023-00013-x.

Epps, Minna. "A Legally Binding Agreement—But What Next?" *IUCN* (2023). https://www.iucn.org/our-union/members/welcome-unite-nature/legally-binding-agreement-what-next.

Marlow, Jeffrey. "The Inside Story of the U.N. High Seas Treaty." *The New Yorker*, March 9, 2023. https://www.newyorker.com/news/daily-comment/the-inside-story-of-the-un-high-seas-treaty.

ternational waters, for over a decade without any success in adopting them. In July 2023, the ISA decided that it was not prepared to issue a mining code, but instead issued a "road map" for reaching consensus on it, with expectations that the regulations would be finalized in July 2024. When debate concluded in Jamaica in August 2024, however, the members walked away without any rules finalized.[18]

This stall in forward progress has a lot to do with the growing chorus of voices calling for a moratorium, precautionary pause, or outright ban on deep-sea mining until scientists can more thoroughly investigate its impacts. At the time of the ISA meeting in the summer of 2024, 32 countries had signed on to the petition for a moratorium, alongside several companies, fishing coalitions, financial institutions, and hundreds of individual scientists and marine policy experts.[19]

Here's the problem that many see with this potential industry: Even as companies actively refine their technological ability to complete deep-sea mining, its impacts on ocean environments, and on people who depend on the oceans, are uncertain. Research has been sparse into how local organisms will react when polymetallic nodules are removed, sediment is kicked up from the seafloor, and unprecedented amounts of noise and light are introduced to this otherwise dark, quiet realm—a habitat whose inhabitants are only partially known to us. Little, too, is known about how these changes will reverberate: how changes in deep-sea populations or pollution from the process, including toxic chemicals discharged in mining wastewater, may filter up through the food chain and into fisheries on which consumers rely.

The problem starts with the fact that scientists simply don't know enough about the abyssal plains in the first place.

"This is one of the most remote ecosystems in the world, so we don't have any long-term time series on it," says Andrew Thaler, a deep-sea ecologist and the former editor of the news site DSM Observer. "For abyssal plains, the mining companies have done a lot of work to fill those gaps, but the gaps remain enormous. We have no baselines, we don't know the natural variability of these ecosystems, and we don't understand how well connected they are across the world."

Scientists do know that polymetallic nodule fields host abundant life down in the darkness. Research on nodule fields in the Clarion-Clipperton Zone, a minerals-rich region between Mexico and Hawaii, found them populated with shrimp, brittle stars and starfish, sea cucumbers, fish, and octopuses. Worms, crustaceans, and other small creatures burrow in the sediment and within living shelters called xenophyophores, ruffled hand-sized creatures that are among the largest single-celled organisms on the planet. The nodules themselves are also a home for life, serving as a foothold for corals, anemones, and sponges, while the holes and cracks on a nodule's surface offer abundant territory where nematodes and microbial life can hide away.[20]

And while these deep-sea worlds might seem distant from our lives here on the surface, there's existing and growing evidence that this habitat is important to humanity. The deep sea is a major piece of the oceanic food chain, and research already suggests that deep-sea mining could disrupt the lives of top-ocean predators, like tuna, on which humanity relies for food.[21] Studying species of seafloor sponges has already yielded a potent form of chemotherapy, the drug eribulin. Other medicines are in the works from the deep sea: a malaria-fighting compound from a sponge living on a seamount in the Pacific; drugs for multidrug-resistant ovarian cancer and leukemia from a feathery sea star relative called a crinoid found off the island of Curaçao; and cancer-killing compounds discovered in a new bacterium that lives at the bottom of the Mariana Trench, just to name a few.[22] These are just the species we've documented; who knows how many more exist in as-yet unexplored regions?

Right now, we have only clues as to what will happen to all this life when mining moves in or what those changes might mean for humans. Exploratory missions have mined nodules in a handful of test sites, including off the coast of Peru, in the waters east of Florida, in the Indian Ocean, and in the Clarion-Clipperton Zone. While patterns differ among these sites, and minor increases have been documented among some species, the consensus of research seems to be that the density and diversity of life decreased at these deep-sea mining sites and that they've so far shown minimal recovery. This isn't particularly surprising to scientists, since nodule fields themselves take millions of years to form. The oldest of these exploratory sites, however, dates back only to the 1970s, and their scale is probably too small to apply the findings to true commercial mining operations.

"It is our view that insufficient information is currently available to generalize the observed biological effects to the longer terms, larger scales, and greater disturbance intensities (e.g., from sediment plumes) expected to result from full-scale mining activities," wrote a group of scientists from across Europe, the United States, and the United Kingdom in their 2017 review of research into biological impacts at all of the known deep-sea mining test sites.[23] They continued: "The experiments that have been carried out are few in number and have been confounded by major differences. . . . In addition, the spatial scales of disturbance (up to tens of square kilometers), and the intensity and duration (a few days) of plume impacts in these experiments are orders of magnitude smaller than will very likely occur for actual mining."

This uncertainty is oddly poetic for the industry. In the spring of 1968, at the height of the Cold War, the CIA searched for a sunken Soviet submarine using the cover story that it was exploring the feasibility of mining polymetallic nodules. Billionaire Howard Hughes provided a smokescreen, accepting a federal contract that made it appear that one of his companies was sending the ship *Hughes Glomar Explorer* to explore this untapped form of mining. The operation, officially called Project Azorian, helped to kick-start interest in the deep-sea mining industry, but it also spawned a famous catchphrase: When the CIA received Freedom of Information Act requests for documents about its actions, the agency responded that it could "neither confirm nor deny" the documents' existence, a line that would become infamous as the "Glomar Response."

Ultimately, the Glomar Response best speaks to the state of the science on deep-sea mining today. We can neither confirm nor deny that deep-sea mining will be sustainable; definitive evidence simply doesn't yet exist.

"There's often this pushback against scientists, saying that we *always* want to know more, asking, when is that knowledge ever going to be enough—but in this case we are still answering the most basic questions," says Diva Amon, a deep-sea biologist and science advisor to the Benioff Ocean Science Laboratory at the University of California–Santa Barbara, whose research focuses in part on the impacts of deep-sea mining. Amon sees the problem with the ISA as putting the cart before the horse: "They've gone ahead and made mining the priority and now retroactively are trying to fit environmental management plans around mining claims. And we are seeing science emerging that says that's not the correct way to do things."

In other words, deep-sea mining faces the same conundrum faced by other forms of ocean energy, like wind and wave-energy technology: The technology is progressing faster than the science around its impacts, as well as governments' ability to develop regulations around how it should be used. But scientists argue that pressing forward in the deep sea is different from building a few pilot wind farms. The fragility of the deep sea, and its importance to ocean life writ large, could mean that small mistakes have huge impacts. And still other layers of nuance have not yet been addressed.

For example, most of the focus on deep-sea mining tends to be on polymetallic nodules, which seem to provide the least destructive form of mining. But two other deep-sea habitats may also be mined under the ISA's permitting systems: cobalt-rich crusts, which occur primarily on underwater mountains called seamounts; and biodiversity-rich hydrothermal vents, which contain a wealth of copper, sulfides, and sometimes gold and silver. Experts have long been concerned that the new ISA regulations do not distinguish these three types of mining and therefore open the door to the destruction of some of the planet's rarest and most valuable ecosystems.

Finally, none of these conversations through the ISA engage with the geopolitical issues around deep-sea mining. One of these is the question of mining in national waters. India, Tonga, the Cook Islands, and Papua New Guinea are already moving forward with plans to mine their national waters, a right enshrined in UNCLOS. Mining companies have also expressed interest in the seabed around US territories in the Western

Deep-Sea Mining Devices

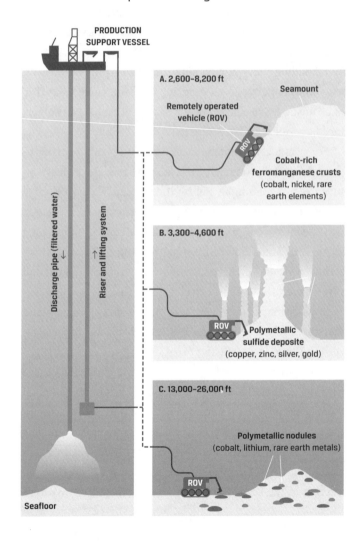

PRODUCTION SUPPORT VESSEL

Discharge pipe (filtered water) ↓

Riser and lifting system ↑

Seafloor

A. 2,600–8,200 ft

Seamount

Remotely operated vehicle (ROV)

ROV

Cobalt-rich ferromanganese crusts (cobalt, nickel, rare earth elements)

B. 3,300–4,600 ft

ROV

Polymetallic sulfide deposite (copper, zinc, silver, gold)

C. 13,000–26,000 ft

ROV

Polymetallic nodules (cobalt, lithium, rare earth metals)

Pacific, although some of this area is protected within the Pacific Remote Islands Marine National Monument, established in 2009 by President George W. Bush, which includes 165 known seamounts. The ISA mining code doesn't legally control any of these kinds of mining, though it is possible that some countries may adopt some of these regulations willingly.

The decision to mine the deep sea is also, for some countries, linked with international relations and issues of national security. In principle, land holds all of the minerals we need for hundreds of years. Why, then, go to the ocean? One reason is because the minerals are sometimes available in higher abundance than on land; for example, copper levels at hydrothermal vent fields can be as much as 10 to 20 times higher than those of the most profitable copper mines on land. But another reason—perhaps the primary motivator for many countries—comes from the pairing of international trade with geopolitics.

Take a country like Japan. Japan produces a lot of advanced electronics, but that production is threatened by China's coupling of rare earth exports to its own political objectives. Japan can instead turn to the ocean as a possible alternate source. This drives activity in their exclusive economic zone, and may lead them to advocate for more industry in international waters.

Today, those in favor of a moratorium argue that the market for these metals is not urgent enough to rush the process of vetting deep-sea mining. After all, even though the necessary technology is available, only a few of the deep-sea vacuums and massive crust-chewing crawlers have been built. Geopolitical developments could create circumstances driving a demand for deep-sea mining, but at present it appears that large-scale activities are decades away. This gives us time to get the science right, but we need to get busy.

Another argument in favor of deep-sea mining is that it could eliminate the negative impacts of mining on land, which so far has a terrible track record; irresponsible practices have long been connected to the destruction of forests, pollution of rivers and coastal waters, and poisoning of drinking water, along with profoundly dangerous labor conditions for some miners. Yet economic analyses suggest that with the growth in this market, terrestrial mining will continue to exist, thanks to the rising demand—and under the ISA's current regulations, countries will not be able to prioritize buying minerals from any one source. It may be more effective to try to ensure that countries with poor environmental records are better-regulated than try to outsource these minerals to a new industry.

Amon has not lost hope for a moratorium. After all, "a few years ago it was almost blasphemy to use the M-word at these meetings," she jokes. "Now, for the first time, during these meetings there is a formal agenda item for discussion of a moratorium. Really, it is a huge shift in the narrative of the discussions that have been happening."

Thaler also notes that deep-sea mining companies and their employees tend to be, on a whole, environmentally minded; in-

POWERED BY THE OCEAN   **143**

deed, this is part of the reason the industry has been in development for nearly two decades without mining a single mineral.

"The reality is that if we didn't care about the environmental impacts, we had the technology in the '70s: You could just drag a trawl across the Pacific and bring up every nodule you want," Thaler says. "The fact that that's not what has happened points to this idea that the industry as a whole is more cautious than terrestrial mining."

Delaying mining to make room for better research—for science that could tell companies where to extract, how much to extract, and where limits need to be placed—offers the deep-sea mining industry an opportunity to set itself apart from all other mining, ensuring that they will steward this ecosystem before removing a single rock from it.

The regulations that go into place to guide this new industry may also reverberate well beyond the deep sea and well beyond our own planet. One of the most remarkable facets of mining for minerals in the deep sea is that the majority of this mining would occur in international waters, where the Law of the Sea dictates that resources are the "common heritage of all mankind." Logistically, this means that the financial benefits of extraction must therefore go to benefit all member nations, a thorny requirement, and one for which the ISA is still creating the appropriate financial structures.

But philosophically, this also means that the value of these resources can't just be defined by the people who see their value in dollars. When a resource belongs to everyone, its value will change depending on that person's perspective. To some, the value of a resource might lie in the intrinsic value of species that depend directly on it; to others, the value might come from the fisheries and wider ocean ecosystems that it benefits, or the tourism that it enables; to others, the value lies in the scientific knowledge it can provide or even in the intangible possibility of future discoveries.

Our ocean provides a clear picture of the delicate balance that hangs between exploration and extraction, between exploitation and understanding. There is still so much to be done to ensure that the ocean we hand down to the next generation is a healthy, biodiverse space that continues to help our planet thrive. But even as we look to the future, and perhaps to a future where we are a species living and operating on multiple worlds, a fascinating symbiosis can happen between ocean science and space exploration. And what we find in extraterrestrial oceans may give us something in return: a new way to view the oceans of our home world and an appreciation of the life that they give to Earth.

CHAPTER 7

Oceans Across the Solar System

ALTHOUGH THEY DIFFER AT THE SURFACE, all latitudes of the world's ocean share one common characteristic: From the tropics to the poles, below around 6,000 feet (2,000 meters) they are uniformly cold, with temperatures that hover somewhere between 40 degrees Fahrenheit (4.5 degrees Celsius) and right around freezing. Which is why, on a February day in 1977, WHOI researchers aboard the RV *Knorr*, led by geologist Robert Ballard, were excited when the instrumented sled they were towing off the Galápagos Islands registered a spike in temperature.[1]

The sled—Acoustically Navigated Geological Underwater Survey, or ANGUS—had been snapping photographs every ten seconds as it was towed along the bottom, so the scientists were eager to haul it back to the surface and review the shots from this hotspot. They paged through slowly. For most of the trip, ANGUS had been capturing a bare seafloor covered in the cracked black bubbles and knobs of cooled lava. But for thirteen frames, right where ANGUS had detected the spike, there was something else: dozens and dozens of bright white clams and brown mussels.

The complement aboard the *Knorr* during that trip was composed entirely of geoscientists. They were there in search of exactly these hotspots, seeking to prove the long-standing hypothesis that water, superheated by the Earth's mantle, spewed out in cracks at certain places along the deep seafloor. But even without a single biologist on board, the team knew that there wasn't supposed to be this much life at such immense depths. Previous expeditions had found only scattered and solitary animals, not dense congregations. Luckily, another Woods Hole ship, the *Lulu*, was operating in the vicinity with a team of biologists, and they had a vehicle aboard that could help verify they were seeing what they thought they were seeing. Moving quickly, the researchers prepped *Alvin* and prepared to dive.

The team of three men descended more than a mile and a half down to the bottom. *Alvin* puttered for some time through the desolate sameness of the seafloor before the sub reached something entirely new. According to accounts of the trip, when they arrived, Oregon State University scientist Jack Corliss communicated up to the surface with a question: "Isn't the deep ocean supposed to be like a desert?" he asked. A graduate student responded in the affirmative, and Corliss continued: "Well, there are all these animals down here."

The geologists knew that miles below them, two continental plates were slowly pulling apart from one another, forming new continental crust in their wake and propelling superheated rock to the surface. But here, the seafloor was covered in cracks, from which issued curtains of shimmering, cloudy blue water: a rain of minerals precipitating out as hot water met the freezing temperatures and rapidly cooled. Surrounding this alien landscape were colonies of foot-long white clams, apparently thriving.

Over a series of dives on *Alvin* through February and March, the team documented an extraordinary variety of life in these crushing depths: brown mussels, blind white crabs, orange siphonophores that swayed above the seafloor like fields of dandelions, a purple octopus, an albino-white skate, and meadows of red-tipped white tube worms that grew like flowers at a site the scientists called "The Garden of Eden." (All of the sites received playful names; the first, clam-filled hotspot was called "Clambake 1.") The expedition quickly burned through their small supply of formaldehyde for collecting and preserving samples and had to switch to vodka instead.[2] It was one of the most extraordinary findings in the history of marine science.[3]

In the nearly five decades since, hydrothermal vents have been found all over world, each flourishing not just with new types of life, but also with a new definition of living. The nearly 60 degrees Fahrenheit vent (15.5 degrees Celsius) temperatures observed on the first *Alvin* dives have been far exceeded: expelled water at the "Two Boats" vents on the Mid-Atlantic Ridge reaches as high as 867 degrees Fahrenheit (464 degrees Celsius), yet the Pompeii worm manages to live below these scalding plumes, exposed to temperatures as high as 212 degrees Fahrenheit (100 degrees Celsius). Within the tissues of many vent organisms, scientists found bacteria that extracted the harsh chemicals spewing from the vents—specifically, hydrogen sulfide, which is toxic to most animals at high doses—and combined it with carbon dioxide and water to produce sugar. These bacteria shared that sugar with their host organisms, like mussels and tube worms. The bacteria were also scavenged by zooplankton, crabs, squat lobsters, and shrimp, which in turn feed fish, eels, octopuses, even sharks—forming the base of an entire ecosystem.

Prior to this discovery, all ecosystems on our planet were thought to be based on photosynthesis; plants and algae make sugars out of sunlight, which feed herbivores, which feed carnivores. Even deep-sea animals that lived out of the reach of the sun, scientists believed, relied in some capacity on food falling from the sunlit world to stay alive. But here was proof of another way to get by. Here was an entire ecosystem instead based off of *chemosynthesis*, where creatures create food by processing chemicals rather than sunlight.

The discovery entirely rewrote our understanding of life on Earth and our theories of where life may have come from. Rather than the warm, shallow pools featured in previous theories, it's now considered possible that hydrothermal vents' hot, mineral-rich waters may have been the nursery for our planet's very first cells. And with that realization came another possibility: If life could survive in the cold and the dark at the bottom of our own ocean, it was possible that we might find the same on oceans elsewhere, off of our home planet—oceans that we now know are spread across the moons of our solar system and through abundant alien worlds beyond.

* * *

There was a time in astronomy, through most of the twentieth century, when Earth felt a bit lonely in the universe. Despite advances like the Hubble Space Telescope, which brought us fantastic images of the universe, our home world was the only one that we knew contained liquid water. Moreover, no one had ever detected a planet outside of our solar system. It seemed possible that our multiplanet system could be an anomaly in the universe, and Earth, the most anomalous planet of all.

Our existential loneliness did not last. The first hints that these assumptions were wrong came in the early 1970s, when models of heat flow and telescope observations suggested that Europa, one of the moons of Jupiter, was covered with ice—and that it was possible there was liquid water beneath it. In 1977, the same year that those out-of-depth geologists discovered life at the Galápagos' hydrothermal vents, NASA launched two probes, *Voyager 1* and *Voyager 2,* to take advantage of the rare alignment of Jupiter, Saturn, Uranus, and Neptune for a flyby mission.

In 1979, images came in from both *Voyager* probes as they flew over Europa's icy surface and found it remarkably free of impact craters, suggesting that the moon's surface was continually renewing and covered in immense cracks that seemed to be filled with a dark, icy material . . . something that looked like it might have once been liquid. (The *Voyager* probes also carried the famous "golden records," created with the help of Carl Sagan, which were made in such a way that theoretically any alien civilization could play them. Included on those records, alongside human languages, music, the din of thunder, trains, and barking dogs, were ocean sounds: crashing waves and whale song from a humpback recorded off Bermuda. *Voyager* 1 and 2 have both now passed out of our solar system and are continuing out into deep space.)

In 1995, despite nearly catastrophic onboard instrument failures, the spacecraft *Galileo* captured even better images of Europa's surface and documented that Jupiter's magnetic field was distorted around the moon. This suggested that an electrically conductive fluid was moving beneath Europa's ice surface and creating a magnetic field particular to the moon itself. This not-so-distant moon seemed to be covered with an ocean of salty liquid water, moving beneath 10–15 miles of ice. Our lonely ocean world suddenly had company.

Europa was far from the last ocean world to join the family. *Galileo* also investigated another of Jupiter's moons, Ganymede, and found evidence that it, too, might be hiding an ocean below its ice. Research since then has confirmed that the saltwater ocean on this moon may be the largest in our solar system.[4] Astronomers had also long known that Saturn's moon Enceladus was covered in ice because of its status as one of the most reflective objects in the solar system; but in 2005, the spacecraft *Cassini* flew through plumes of vapor spewing from Enceladus's southern pole, finding evidence of salty liquid water there beneath the bright cold.[5] More recent analysis of *Cassini* data found that the plumes from this pole contained not only sodium, potassium, chlorine, and carbonate—ingredients that could make this ocean habitable—but also phosphate, a much rarer ingredient in the universe and one of the essential building blocks for DNA.[6] In December 2023, researchers announced that these plumes also included particles of hydrogen cyanide, a toxic gas that nonetheless may be the starting point of many life-bearing chemical reactions.[7]

The family continues to grow as we learn more about our home system. Jupiter's Callisto may have a liquid ocean beneath a crust of rock; Triton, orbiting Neptune, may have a nitrogen-rich sea beneath its icy surface; and Ceres, a dwarf planet between Mars and Jupiter, appears to have briny underground lakes and ice-spewing cryovolcanoes. The surface of Titan, Saturn's largest moon, is dotted with lakes and rivers of liquid methane and ethane, as well as wind, rain, and seasonal weather patterns just like Earth's; in fact, scientists see the moon as a potential analog for our primordial planet, containing all the same building blocks for life.[8] The low density of Saturn's moon Mimas (dubbed the "Death Star moon" for its oddly close resemblance to the fictional Star Wars weapon) suggests that it's made mostly of water ice, and it too may host a liquid ocean underneath.[9] And the mysterious, hundred-mile-long fault lines across Pluto suggest that even our system's most distant planetary body might also have liquid oceans beneath its plains of nitrogen ice.[10]

"I believe that if we're going to find signs of life elsewhere in the solar system, we're actually looking in the wrong place," says Robert Braun, head of the Johns Hopkins Applied Physics Laboratory's (APL) Space Exploration Sector. "We've spent a lot of time looking at Mars, which was once a warm wet world, but that was 3.5 billion years ago. Whereas you go out to Jupiter or Saturn, there is abundant water and abundant energy. They're essentially their own mini–solar systems."

What's more, we now know that our solar system's planetary tendency is far from an anomaly—and that our neighboring planets are, in fact, just a handful in a vast extended interstellar family.

In 1992, astronomers used some clever math to calculate that there were two planets orbiting a neutron star some 2,300 light-years away—the first exoplanets ever discovered. This star would constantly be bathing its planets in radiation, potentially making them uninhabitable. Another spark came in 1995: Two researchers found a planet half the size of Jupiter closely orbiting the star 51 Pegasi, 50 light-years from Earth.

They detected the distant world by examining how the planet made its star wobble very slightly in its orbit.

The floodgates did not truly open until astronomers refined a technique called the transit method: a way of spotting exoplanets by detecting the minuscule decrease in light that happens when a planet crosses in front of its home star. In 1999, using this method, two teams independently found a total of three planets in the Pegasus constellation. Since then, the launch of specialized telescopes and new missions targeted at finding exoplanets has sent the number of detected distant worlds skyrocketing. NASA's exoplanet catalog currently hosts over 5,600 confirmed discoveries, most of which were found with the transit method.[11]

Scientists then observed that by analyzing the spectrum of light that passed through these planets' atmospheres, they could learn what that atmosphere was made of. Over the years, keen eyes discovered many planets that orbited their sun within the habitable zone, also called the "Goldilocks zone," where it's not too hot, but not too cold: in other words, just warm enough that water could stay liquid on the surface or beneath ice. But it wasn't until 2019 that researchers at University College London, using data from the Hubble Space Telescope, identified a planet in the habitable zone that contained water vapor in its atmosphere.

The planet, K2-18b, spins in space about 110 light-years away. It appears to have a hydrogen atmosphere and a density somewhere between that of Earth and Neptune. It's a tantalizing discovery: That density could mean that K2-18b is rocky beneath its thick atmosphere—but it also could be composed almost entirely of gases and liquids. If it is rocky, that might indicate that it has liquid water and oceans on its surface. K2-18b's star is a red dwarf, dimmer than our sun, but the planet is closer to its star and so receives about the same amount of light that we do. The only difference is that the red dwarf may bathe K2-18b's surface with more high-energy radiation, tough conditions for any creature to survive. Yet as we know well from our own oceans, life often finds a way.

If there is life out there in the blackness, how will we find it? The distance to these moons and distant planets seems daunting, but with the perspective of history, we're nearly at the same place that explorers on the *Challenger* expedition were in the 1870s, faced with the first conclusive traces of life in the deepest oceans. It was just over a century later that researchers were peering through the tiny portholes on *Alvin* and looking at the profusion of life two miles down, crowded around hydrothermal vents.

Indeed, NASA is looking to our ocean as one of the best places to prepare for what might lie ahead, when our robots reach these seas out in space. Many marine habitats offer helpful analogs for conditions on moons like Europa or Enceladus, two of the top candidates for upcoming missions, and so provide good environments to test the science and technology that will explore them. Experts in marine ecosystems and biogeochemistry can help set the parameters for these missions, providing an idea of the most likely conditions in which that life might form. And researchers versed in detecting signs of life in some of Earth's most inhospitable environments, such as its deep-sea trenches or coldest ice-capped seas, can give us signals to look for to prove that life is really there.

GOING DEEP

If liquid water exists on Europa and Enceladus—and the evidence for it is strong, though it still needs confirmation on future missions—then the water there starts deep. The ice on Enceladus may have shallow points at its south pole, where its ice is only about 3 miles (5 km) thick, but across most of the planet the ice thickness likely extends between 12 and 16 miles (19–26 km) over a 25-mile (40 km)-deep ocean. Europa contains no known shallow spots, and the moon is likely covered with 10 to 15 miles (16–24 km) of ice before reaching liquid water. The ocean itself then reaches down another estimated 40 to 100 miles (64–161 km).

Any robotic systems exploring beneath the crusts of these icy moons would do so under extraordinary water pressures. Our future explorers enjoy one benefit: Because these moons have a fraction of the gravity of Earth, this pressure will not be nearly as intense as it would be under the same depths on Earth. On Europa, for example, the pressure at the seafloor would be between about 19,000 and 38,000 pounds (8,600–17, 236 kilograms) per square inch (psi).[12] The pressure at the bottom of the seven-mile-deep Mariana Trench, the deepest part of Earth's seafloor, reaches 15,750 pounds (7,144 kilograms) psi. Since researchers have successfully explored there with ROVs and with crewed submersibles, this puts Europa's seafloor within reach of technology we already have, and Enceladus well in range: Its diameter is six times smaller than that of Europa, so it would have even less gravity.

With all of that in mind, the deepest part of our ocean serves as a good test bed for the technology that will one day plunge beneath the icy crusts of these ocean worlds. These deepest parts of our ocean are called the hadal zone, named for Hades, the Greek god of the underworld. Appropriately, when WHOI and NASA's Jet Propulsion Laboratory built an AUV to test some of the technology they might one day send to ocean worlds in space, they called it *Orpheus*, after the Greek hero who ventured into the underworld and returned home to tell the tale. *Orpheus* is equipped with sensitive cameras and a visual navigation and mapping system, similar to the technology used by NASA's *Perseverance* Mars Rover. This system creates three-dimensional maps ranging around shells, corals, and rock formations that serve as landmarks (or perhaps more appropriately, seamarks) that the robot uses to orient itself and recognize places where it's already been. This is especially useful in the dark and often murky depths of the deep sea, but it could also enable a vehicle to navigate unknown extraterrestrial oceans with features and visibility that we may not be able to predict until we get there.

* * *

Darlene Lim, a NASA planetary geologist, was serving on the NOAA Ocean Exploration Advisory Board when she got the idea that it might be useful for the ocean and space exploration communities to learn from each other.

"It became evident to me that the way that the science and ocean exploration community has organized itself—to go out to sea, and also enable those who cannot go out to sea to have access to that data—could help to enable a science team on Earth to be much more present and integrated in mission operations when we go to the moon or onwards," Lim says. Today, she is the principal investigator of NASA's Systematic Underwater Biogeochemical Science and Exploration Analog, also known as SUBSEA, a program working to bring ocean exploration and space exploration together for the benefit of both.

In 2018 and 2019, SUBSEA carried out two missions to hydrothermal vents in the Pacific in partnership with the Ocean Exploration Trust, where they tested not only the technology that they might send to ocean worlds, but also what it might be like for people to work on these missions from millions of miles away. The target was Kamaʻehuakanaloa, formerly called the Lōʻihi seamount, an active undersea volcano southeast of the island of

Hawaii. While this volcano isn't erupting at present, it's covered in hydrothermal vents; but unlike the high-temperature chimney vents that you might find in the deep sea, the water coming out of these vents is at a relatively low temperature—less than 212 degrees Fahrenheit (100 degrees Celsius)—and the vents' shallow depths mean that the water pressure is much lower, so hot water drifts rather than jets out. This makes Kamaʻehuakanaloa a good analog for vents at Enceladus's shallow south pole—and, Lim says, simply finding these hard-to-locate vents is a valuable exercise in what it might be like to explore the icy moon's seafloor, where scientists are uncertain how active the geology will be.[13]

While exploring Kamaʻehuakanaloa, the SUBSEA team tested an underwater drill that could be used both for oceanographic missions and eventually in space; it clamps onto rock to compensate for the low gravity. The team took samples to look for hydrogen-eating microbes, which could be an analog for life on hydrogen-rich Enceladus. And on this mission and a subsequent SUBSEA trip off the coast of Oregon, they tested what it would be like for scientists back home to manage mission parameters and observe dives remotely, through telepresence.

"There was a wonderful confluence of interest in that natural science team, in wanting to explore this planet, and also as a springboard of knowledge to understand other ocean environments," Lim says.

With the help of social scientist Zara Mirmalek, the SUBSEA team also used data from the 2018 and 2019 expeditions to develop a simulation that allowed NASA team members to complete ocean exploration operations that more closely resembled how they would carry out operations in space. These elements ranged from simple adjustments, such as incorporating a brief time lag to simulate the transmission delays of space, to comprehensive models, like a predeparture system in which all scientists involved outlined the specific operations needed for their research—anticipating the sort of missions where it wouldn't be possible to change the program once their robot was sitting on a different celestial body.

"We created a structure that better enabled ocean scientists onshore to be participatory in real-time operations on the boat," Lim says. That structure became especially important to the Ocean Exploration Trust when the COVID-19 pandemic nearly shut down all at-sea operations. "Suddenly they had a structure and a foundation to pull from to go to sea with a skeleton crew and still be able to have science done onshore that represented many scientific interests," Lim explained.

Before scientists remotely direct a mission to Europa or Enceladus, they're focusing on operations a bit closer to home. One of Lim's projects has been collaborating on a rover expedition to map ice on our own moon. This will be an essential step toward establishing a more permanent human presence on our nearest celestial satellite—and creating a place where missions heading further into the solar system can stop to fuel up or collect other important resources.

Unlike Mars or further targets in our solar system, the Moon is close enough that there is a negligible communications delay for vehicles on its surface; about 1.28 seconds for radio transmission to travel each way, faster than the roughly 4 second one-way acoustic transmission time to a rover on the seafloor at a 20,000-foot (6,100 meter) depth. Using the telepresence lessons they took from SUBSEA, Lim and her team have created a mission science center that will fully integrate scientific operations into the broader moon mission. When this mission launches, for the first time in NASA history science decisions won't be happening in a back room and then brought to mission control, Lim says; researchers will instead be a part of all major critical mission planning, incorporating science into this first step toward extending humanity's reach farther into the solar system.

THINKING SMALL

In the annals of science fiction, Europa looms large as a place where life may not just be present but potentially may be multicellular, complex, and even intelligent. Arthur C. Clarke's *2010: Odyssey Two*, a sequel to the famous *2001: A Space Odyssey*, imagines Europa as a place full of intelligent aquatic life with such evolutionary potential that aliens deem that it must be protected from humanity and send Earth the message "ALL THESE WORLDS ARE YOURS—EXCEPT EUROPA. ATTEMPT NO LANDING THERE." The movie *Europa Report* and the video game Barotrauma depict the moon as populated by animals that fight back against humans' arrival, whether with tentacles or toothy mouths.

Perhaps these storytelling dreamers will be right (it wouldn't be the first time). But if there is any life in our solar system's ice-locked oceans, researchers' bets are currently that it will be microbial. Microbes are, after all, much more pervasive on Earth in both space and time; single-celled organisms appeared on our planet about 4 billion years ago, not long after our oceans formed, but multicellular life didn't show up until around 1.5 billion years ago. Those eons of dominance can be seen in how bacteria still largely outnumber more "complex" life forms: Measured in gigatons of carbon (Gt C), bacteria are second by weight on our planet only to plants, weighing in at an estimated 70 Gt C. All animals together take up 2 Gt C, and humans, just 0.06—less than 1 percent of life.[14]

The best estimates of Enceladus suggest that its ocean is about 1 billion years old, putting it early in the potential timeline for life. Europa's ocean could be as old as the moon itself, indeed about as old as Earth's, at about 4 billion years old.[15] It remains to be seen whether life would follow the same timescales in a sunless, ice-covered sea as it did in the comparably friendly circumstances of our home world.

What scientists do know is that if there *are* extraterrestrial organisms in these places, detecting them will be a lot harder than turning on a camera and waiting for something to paddle by. As a result, missions to ocean worlds are taking lessons from how marine microbiologists detect life in Earth's deep sea.

Thanks to the ice shells of most extraterrestrial ocean worlds, "their ocean is not going to work the way that ours is, with energy from the sun driving the microbial food web from the surface," says Julie Huber, a senior scientist at WHOI who studies microbes that live below the seafloor, in cracks created by hydrothermal activity—and who also works with colleagues in astronomy to help them plan missions to ocean worlds. "Studying microbes from the deep ocean and hydrothermal vents might be a way to imagine how these ocean worlds are functioning."

One way that microbial oceanography is helpful is in revealing the types of environments that these microbes favor. For example, Huber says, "hydrogen is absolutely delicious from a microbial perspective: It has a really high energy yield, and if there's any high energy around, they will use it if they can." The *Cassini* mission's flyby of Enceladus detected hydrogen in the plumes, and hydrogen in a molecular form (H_2) that's familiar from Earth: It forms when water interacts with the crust around hydrothermal vents. This tells astronomers to pack their probes' proverbial suitcases with tools that can perform Raman spectroscopy, a type of detailed chemical analysis that can detect molecules like H_2, and point them toward places where hydrogen-snacking organisms might live.

Huber is also talking to mission planners about the byproducts that microbes produce on Earth, and what would be realistic to look for as a signal of life. She gave the example of a recent conversation in which an astronomy colleague suggested detecting the intermediate chemicals produced through metabolism, when microbes process sugars to create energy; in other words, chemicals that, on Earth, we *only* see created by living things. Huber and another microbiologist had to explain that these byproducts are usually consumed as rapidly as they're created and so don't usually persist long enough to be detectable.

"It's often me digging into the literature trying to help them understand here's what we have measured on Earth, this is the detection limit we're comfortable with, or, here's why this won't be a useful measurement," she says.

That question of a "detection limit" is one that haunts Huber: It's the term for the minimum amount of a molecule or compound that microbiologists are comfortable saying counts as a sign of organic life. "My big fear is we're going to launch a mission and people aren't going to believe the result because the technology wasn't vetted enough," Huber says. That's why she advocates for multiple lines of evidence that indicate signatures of life: compounds like formate and ATP, which are produced by microbial metabolism, but also information-carrying molecules like amino acids or even DNA and RNA. "Exceptional claims require exceptional evidence," she adds, voicing a principle deeply embedded in the workings of science.

After decades of science fiction depicting alien life as complex, animalistic, and often intelligent beings, there's some sense that finding nothing more than cells on an ocean world might, to some, be a letdown. But Huber doesn't see it that way, and not just because she finds microbial life fascinating.

"I think it would be so profound, because it would show, especially in these outer moons, that there is another genesis of life in our solar system," she says. "This is where life started on this planet, and I think it's really fascinating to consider where are these other ocean worlds on their evolutionary trajectory."

Such a discovery could also help us reflect on how our own ocean, our own complex biosphere, got to where it is today. After all, most of what we know about life is based on what has survived on Earth to be studied; a new data point on how life can evolve might show us the potential pathways that are out there, and perhaps expand our definition of life itself.

CHAPTER 8

Could Deep-Sea Robots Discover Life in Deep Space?

UNTIL RECENTLY, THE PROCESS of most ocean research demanded patience. If your research relied on data gathered from out on a ship, months or years might elapse between periods of data-gathering. Each trip required the researcher, or their lab members, to collect the data they needed manually, in person. In the intervening time, there was always some risk that the object of your study might have changed dramatically or have disappeared altogether.

The advent of remote-sensing technologies has changed how scientists studying the ocean conduct research. Buoys, moorings, floats, and satellites allow us to keep a constant eye on many parts of the ocean. We can measure enormous amounts of basic data that enable us to understand our world, and increasingly, we can capture more complex information like visual data, sound, and biological activity. AUVs and ROVs enable researchers to gather the information they need spatially, in a single trip, or over time, as the robot spends months at sea while the researcher is comfortably back at home. In brief, the tempo of ocean science is accelerating.

But the same isn't true for space. Each visit to a moon of Jupiter or Saturn must traverse vast distances, and the immense challenges in getting our technology there mean that the time period between each voyage of discovery is long. Nor is it possible to build one 'super spacecraft' that can make all the desired measurements in one mission. Each mission builds on the last. Scientists do not know the next set of questions until they've answered or partially answered the first. Consequently, more than twenty years elapsed between the tantalizing images of Europa taken by the *Galileo* probe and the most recent images, snapped in 2022 by the *Juno* spacecraft during a flyby. *Voyager 1* took only a year and a half to reach Jupiter, but it was moving quickly—a so-called flyby mission—because it wasn't designed to stay and study the moons and planets that it passed; *Juno,* by contrast, took more than four years to make the same trip, moving at a rate slow enough that it could be inserted into Jupiter's orbit.

Oddly, the challenge of sending spacecraft to the outer solar system is less about accelerating them to high speeds and much more about stopping when they get there; the gas giants Jupiter, Saturn, Neptune, and Uranus have enormous gravitational pull, creating "gravity wells" into which an arriving spacecraft falls, accelerating as it approaches. Depending on the angle, a spacecraft might use that pull to slingshot around the planet, as did the *Voyagers*. However, if you want your spacecraft to stay, you have to slow it down, which requires fuel, which in turn adds weight—weight that could be science sensors. So it makes sense to take a little longer to arrive to maximize the scientific capabilities of the spacecraft.

We could send more probes, to be sure. Perhaps one day the cost of space missions will diminish enough, and the logistics of space travel become sufficiently routine that we can do so. Currently, however, the cost of outer-planetary missions tends to run to billions of dollars, and they often exceed their budgets. (*Europa Clipper*, launched on October 14, 2024, will cost an estimated $5 billion—and take five and a half years to arrive at its destination.) Although reusable rockets are making trips to space much more routine, sending lots of advanced spacecraft to the outer solar system isn't yet financially feasible, either for national space programs or for private entities.

So how can scientists make astronomy more like oceanography? Getting there less expensively and more often, yes, but also, getting the most bang for your buck from every minute of observation time you have, without losing time to errors, damage, or waiting for instructions to arrive from Earth. To do that, the spacecraft and the robots that may find life outside of our home world will need to be smarter than those we have sent before. They'll need to be capable of repairing themselves, adapting on the fly, and making decisions about what data are most relevant to collect. Like many science fiction futures, this capability is already well on its way to becoming a reality—and in some cases, is already being put to use.

LEARNING FROM THE PAST

For humans and robots alike, space isn't the friendliest place to go exploring. Unlike ocean research, space won't crush your vehicle like a tin can, and living organisms won't try to grow on your sensors. Yet those small mercies are more than counterbalanced by intense cold and heat, powerful radiation that can fry electronics, and the potential physical damage from micrometeorites or much larger space rocks. Over time, a vehicle can run out of fuel, get lost, lose communications with home, suffer radiation damage, or even just get too old to go on. And all of that is before landing on any moon or planet. Assuming that our probes make it to another world, they'll then be facing a whole new set of environmental factors that could spell trouble for an explorer of alien planets.

These potential problems are far from theoretical. Space agencies have a long history of facing unexpected roadblocks and coming up with on-the-spot fixes to compensate for them. Among the most notable is the failure of an antenna from the *Galileo* probe in 1991. This antenna was the high-gain transmitter, meaning it could quickly transmit the large amounts of data that it was gathering back to Earth. It was constructed by engineers to unfold like an umbrella, but when home sent the order to open, one of the antenna's ribs jammed, leaving it half unfurled.

Mission control tried swinging the spacecraft toward and away from the sun, hoping that expansion and contraction from the temperature change might wiggle the stuck rib loose; no luck. They tried, six times, rotating the spacecraft and making it come to a hard stop to shake the rib loose, then pulsing the unfurling motors to send vibrations through it. The rib was still stuck. Scientists eventually concluded that four years of storage (caused by the 1986 explosion of the space shuttle *Challenger*, which delayed *Galileo's* launch), and the vibration of three cross-country truck trips had eroded the lubricants in the ribs. Researchers were forced to carry out the probe's science mission using its low-gain antenna, which could send only a fraction of the data that had been planned.

Ever since *Galileo*, when teams are designing vehicles and instruments for deep space, "one of the big variables is that we try to avoid moving parts," says Louise Prockter, former chief scientist for the space exploration sector at Johns Hopkins' Applied Physics Laboratory (APL), who spent a decade as the deputy project scientist on *Europa Clipper* and then served as the co-investigator of the vehicle's camera team. As someone with a particular interest in Europa and in imaging, Prockter is intimately aware of how one failure can set science back: Although *Galileo* still gathered invaluable information about the icy moon, it was only able to take high-resolution images of two pole-to-pole stripes of Europa's surface; anything larger would have taken more bandwidth to transmit.

"I've been doing science on the same images, the same small number of images, we got from *Galileo* for twenty years now," Prockter said. "I personally am very excited about the fact that we're going to get very high-resolution images [from *Europa Clipper*] of pretty much the whole surface."

These past failures factor significantly in how engineers prepare for future missions, and how they ensure that they will get the most bang for every buck spent—something you can see clearly in the design of *Europa Clipper*. NASA's teams now design camera equipment for their spacecraft so that if the gimbal rotating the camera fails, it will fail in a downward position and therefore still be capable of collecting imagery. A similar principle went into the design of *Clipper*'s enormous solar panels. (If stacked on top of each other, these panels would be as tall as the Statue of Liberty.) Though the panels themselves are complex instruments, their deployment was simple, intended to avoid moving parts that could get stuck in transit: An electrical current burned through the wire attaching them to the body of the spacecraft, which then rotated slowly, flinging the solar "wings" out to unfold by centrifugal force.

When *Cassini* flew by the Jupiter system, its radar was mounted on the opposite side of the spacecraft from its camera; it couldn't use both simultaneously, leaving scientists with no way to compare different types of data for the same point at the same time. In contrast, *Europa Clipper* will be able to use its nine science instruments plus the radio communications onboard all at once, so that scientists can capture visual, composition, subsurface, and radiation data simultaneously.

"When we started planning for this mission many years ago, we knew right up front: The instruments need to be on in the same place and same environment," Prockter said. "For example, if there are plumes coming off of Europa, we're going to want to look at them with everything at our disposal, not wait a few more orbits and come back and see that they're not going off any more."

Jenny Kampmeier, a science systems engineer on *Clipper*, adds that the multiple instruments not only provide redundancies but can also help build confidence in their findings. Some science wouldn't even be possible without those redundancies. "In some cases, things have complementary and complex interactions," she said. "The plasma field affects the magnetic field, which affects the electric currents on the planet—if we were only measuring one, we would have too many variables, which means we can't solve this equation."

Finally, because of the immense distances between the Jupiter system and home, it takes forty-five minutes for a message to travel in either direction. That means an hour and a half total for mission control to send a command and receive confirmation that their command went through. As a result, autonomy will be an essential piece of *Clipper*'s mission: Managers back on Earth will send *Clipper* the plans for its science objectives, and the spacecraft will then execute them on its own.

Part of *Clipper*'s autonomy is also focused on protecting itself when something has gone wrong. Kampmeier gave the example of radiation, which is so intense around Jupiter that every day, Europa's surface receives about 1,800 times the average annual dose someone at sea level on Earth receives in a year. This intense bombardment could cause a "bit flip," in which the binary code used to store data on the *Clipper*'s computer inadvertently flipped from a 0 to a 1, or vice versa.[1] This bit flip could be harmless, but it could also be catastrophic, causing an instrument to start filling up its data recorder with garbage data.

Clipper is programmed with a system that recognizes when something like this happens, through a model that monitors all sorts of parameters: voltage, temperature, data production, radiation levels, data storage levels, and hardware states. When something changes in a way that's different from past states, the model recognizes it, and the spacecraft can react accordingly. This ranges from asking home for instructions to flipping on a cooling or heating system, all the way to putting itself into safe mode: All instrumentation will shut down, and the craft will move to a stable configuration, turning to make sure that its antenna is pointed toward Earth and its solar panels toward the sun. There, it waits for instructions from home on what to do next.

The level of autonomy on *Clipper*, though certainly complex, is deliberately conservative compared with, say, the technology regularly used in vehicles that explore our oceans or the sorts being developed by some Silicon Valley tech companies. NASA favors technology that is known to be reliable on big, complex missions like this one and shies away from "technology for technology's sake." This doesn't mean that autonomy isn't playing a role in how NASA is exploring the solar system. Other efforts have already begun to push the boundaries on who makes decisions during missions, moving some of that control out of human hands and over to computers.

DECISION-MAKING ON MARS

When the Mars 2020 mission touched down on the red planet in February 2021, carrying the *Perseverance*, Mars rover scientists knew that its landing would be hectic. Mars's atmosphere is much thinner than Earth's, meaning that there's less of a cushion to slow a landing. But a Mars craft can't fire downward-facing retrorockets to slow down, like astronauts did when they landed on the atmosphere-free Moon; doing so could create enough turbulence to rip the vehicle apart. And because of the twenty-minute communications delay across the distance between Earth and Mars, the control room can't send directions mid-entry if something goes wrong. Engineers must simply trust that everything they've spent years planning will work as it's supposed to.

From entering Mars's atmosphere to landing on its surface takes seven minutes; it's known as "seven minutes of terror."

On top of that, the Mars 2020 mission sought to land in a place on the planet that previous missions wouldn't have attempted exploring: Jezero Crater, an ancient dried lakebed, where harsher terrain increased the odds that the vehicle might

Getting to Europa's Oceans

When *Europa Clipper* slips into orbit around the icy shell of Jupiter's sixth moon, one of the many variables it will scan for on the surface is a good landing point. *Clipper* will never touch down on Europa, but it will be helping scientists plan for future missions that will explore the moon's icy surface and, one day, plunge beneath the ice in search of life in these frozen waters.

However, penetrating this ice—and, eventually, getting *beneath* it— represents a huge challenge for missions to come.

The first mission that will attempt to sample the ice of another world will be the *Europa Lander,* a NASA mission concept currently slated for the 2030s. *Europa Lander*'s goal will be to look directly for biosignatures: traces of life that, over centuries, may have worked their way up from the moon's deep ocean to the surface ice. The goal is to drill down 4 inches (10 cm) or deeper, where any biological material would be shielded from the intense radiation that's constantly bombarding the moon from Jupiter. At the time of writing, the lander's project team was still developing the instrument that will cut into the ice and gather samples. But early tests by the lander's sampling team suggested that they may use a blade that simultaneously scrapes at the ice and fractures it. This is a much more power-efficient way to gather a sample, and the lander's battery will be very carefully rationed.

The goal, of course, is to one day get to the ocean below the ice. On Europa, that ice is estimated to be 10 to 15 miles (16–24 km) thick—

two to five times deeper than the East Antarctic Ice Sheet, the thickest ice on Earth, and well past the depth that Earth-bound ice drills can currently penetrate. The deepest that humans have drilled in ice is 7,060 feet (2,152 meters), a little over 1.3 miles (2 km).

Penetrating ice using only mechanical force, spinning downward like the drill you might use to put a screw into a wall, requires a lot of power. So many of the probes and ice drills being planned for icy ocean worlds will use heat, usually generated by nuclear power. Some, called cryobots, melt through the ice passively, simply falling downward under the force of their own weight. Others will use heat to make it easier for a mechanical drill to do its work, which would also make the probe capable of adapting if it encounters a rock or some other obstacle as it descends.

Planning for an ice-penetrating mission hasn't formally started at the world's space agencies, but NASA, ESA, and other research organizations are already at work testing drill prototypes. One of the most advanced is called VALKYRIE, developed by a private company called Stone Aerospace. In contrast to other designs, this cryobot uses either a heated nose to passively fall through the ice, or a laser-heated jet of hot water to melt the path ahead, relying on a radar system to navigate around obstacles along the way.

After successful tests on the Matanuska Glacier in Alaska, Stone integrated this technology into a new prototype, known as SPINDLE, that combines the heat-powered melt probe with an AUV. SPINDLE would melt its way through the ice using a laser, potentially dropping radio beacons along the way to provide a chain of communications and then deploy its AUV once the probe emerges into liquid water. There, the submersible will be able to roam for up to a mile from the probe to gather water samples and other data. After each mission, it will return to the probe in order to upload data and beam it back to Earth, to tell us what it has found swimming in these faraway seas.

FURTHER READING

"Europa Lander." NASA Jet Propulsion Laboratory. https://www.jpl.nasa.gov/missions/europa-lander.

Oberhaus, Daniel. "An Alien-Hunting Submarine Is Being Tested in Antarctica." *VICE Motherboard,* May 7, 2017. https://www.vice.com/en/article/jpym5x/alien-hunting-sub-europa-artemis-stone-aerospace.

"VALKYRIE." Stone Aerospace. https://stoneaerospace.com/valkyrie/.

encounter large boulders, uneven ground, even steep cliffs. Because of that time delay, the control room crew back on Earth knew they wouldn't have time to adapt to the conditions on the ground below as the craft headed toward the surface, so they let the computer decide.

When the seven minutes of terror arrived, the Mars module made it safely through the Martian atmosphere. But as it floated over the dry red dirt on its parachute, the wind blew it southeast of the flat, safe landing site the planning team had intended. So the vehicle's Terrain Relative Navigation (TRN) system referred to the map its creators had provided, labeled with known safe areas. Based on the amount of fuel left, the computer selected a nice, flat place without too many large boulders to settle down and start its mission.[2]

Giving a spacecraft control over where it lands was a move that NASA wasn't prepared to make for a long time: "The mission managers are very conscious of risk, and if they don't need to use a new technology to accomplish a mission, they won't," says Richard Doyle, former manager of the Information Technologies and Software Systems Division at NASA's Jet Propulsion Laboratory.

But Doyle says that in the case of Mars 2020, decades of technological development in computer vision lined up with a compelling science mission. Some 3.5 billion years ago, a river spilled over the edges of Jezero Crater and carried minerals into the former lake that could have supported microbial life.

Perseverance's mission includes searching for evidence of that past life. The geological layers along the crater's edge, too, may provide a glimpse into the different eras of Mars's history. The potential for new discoveries outweighed the risk of putting the computer in control, and the mission managers decided to give TRN a try. And when it succeeded, TRN became what the industry calls "flight proven," entering the pool of potential tools that NASA is already considering for future, even more dangerous missions.

Unlike the Moon and Mars, which are within range of Earth's telescopes, there aren't any high-resolution images of the surface of Europa, making landing there even trickier. Although *Europa Clipper* will change that for some destinations, one spacecraft can't cover everything. As a result, NASA plans to hand some of the decision-making for an eventual Europa lander to TRN and to other intelligent landing systems. This will compensate for the fact that the terrain could be even harsher than expected and the resources for the spacecraft even more limited.[3]

A similar system will also guide the flight of another spacecraft, called *Dragonfly*, being developed at APL to explore the nitrogen-rich skies of Titan, our early-Earth analog orbiting Saturn. This car-sized, helicopter-like drone—scheduled to launch in 2028, and to reach Titan by 2034—will use a form of autonomy to find a safe spot to land and then use that as a home base as it flies out on scouting missions. Along the way, a system like TRN would then locate safe new landing spots for *Dragonfly* to touch down as it leapfrogs more than 108 miles (175 kilometers) across the moon's surface, stopping to sample the diverse dunes, lakes, and rivers along the way.

Doyle sees a similar shift ahead when it comes to giving computers some control over data-gathering for science missions. The proof that this was possible also came from Mars, as did the opportunity to study a phenomenon that researchers never thought they could study in real time: dust devils, whirlwinds across the Martian surface that pop up unexpectedly and disappear just as quickly. These miniature dust storms can affect operations on the ground; they also make up part of Mars's climate models, of interest both to mission planning and to understanding the planet itself. However, they're too ephemeral and unpredictable to be simulated. The only way to adequately study them would be to do so in person—or whatever the robotic equivalent might be.

But previously, researchers only knew that a dust devil had happened after they downloaded days of data and saw the storm had been there and gone. "For a long time, scientists assumed that they wouldn't be able to observe any dynamic and transient kind of phenomena in real time because of the time delay," Doyle said. "By the time scientists on the ground realized something interesting was unfolding on Mars, that's come out in data that was hours or even days ago."

Artificial intelligence software changed that. By equipping Mars rovers with software that could recognize dust devils and decide what data to collect, scientists today receive regular data packages any time *Perseverance* has a chance encounter with one of these whirlwinds. These packets can include a video of the storm, tracking information on its movement and other data such as wind speed, temperature, dust levels, air pressure, and even the storm's rumble as it moves past. *Perseverance* can even decide to interrupt other mission tasks, such as transiting between points, in order to gather such data.

Although weather observation might seem like a fairly minor use of computer intelligence, Doyle sees it as a good proving ground for software that might one day be used to detect life on worlds too far away for humans to realistically control a probe. "The success with the dust devil detector kind of opens up thinking to what the possibilities are, and more scientists are in that conversation than before," says Doyle.

One clear potential application of this software is with the water vapor plumes that scientists believe may spout from some places on the surface of both Europa and Enceladus. Just like with dust devils, Doyle imagines that artificial intelligence could be invaluable in gathering data on these kinds of dynamic, ephemeral events: "If you can detect it while it's happening, and study it while it's happening, that would be enormous," he said. That doesn't mean that humans will be entirely cut out of the picture, however; scientists will need to help design and update these programs, Doyle says, telling the computer what it should do when it spots something of interest. "The robots are like their proxy, so you need to ask if the scientists were there, what they would want to attend to?"

Systems like these will be game-changing for missions to ocean worlds, but they also operate off of parameters that humans set ahead of time. NASA knew that its Mars rover might encounter dust devils, so scientists told it to look for dust devils. The next challenge comes from preparing our robotic proxies to enter into environments that are largely unknown—places like the ocean below the ice on Europa or Enceladus, where we can only make educated guesses about what they might encounter. Machine-learning experts are beginning to push the field toward systems that can decide to study something even if they don't know what it is and systems that can adapt to challenges or obstacles even when they can't phone home for help. Perhaps unsurprisingly, systems that can do just that are coming out of work already happening on AUVs in Earth's own oceans.

PUTTING CURIOSITY INTO CODE

Sometime around 10 p.m. on Thursday, September 12, 2013, a *Tethys* LRAUV we were running in Monterey Bay, California, got itself into trouble. These vehicles are designed so that some of their contents can slide around within the body, using the mass of the batteries and other electronic components as a weight that helps the robot dive. But during the *Tethys*'s descent that night, the mechanism that shifted the battery back and forth—using that weight to more efficiently control its dives—failed. Because the vehicle was heading down, the battery slid all the way forward, holding the vehicle's nose firmly pointed toward the bottom.

The failure stemmed from an earlier incident: *Tethys* had been attacked by a shark on another dive, which shook the vehicle so vigorously that it left teeth embedded in *Tethys*'s structure. But unfortunately, we didn't know about the internal damage until the weight shift mechanism failed on the 12th. Around 11:15, the vehicle's software identified the problem as a "failure to ascend" and triggered the safety behaviors, which normally would have made the vehicle automatically return to the surface. The problem: The recovery methods programmed into the vehicle all hinged on continuing to drive the vehicle forward. In this case, driving forward simply carried the nose-heavy AUV deeper into the seafloor. Back home, all we knew was that we had lost contact.

Ultimately, this *Tethys* vehicle ended up on the seafloor for twenty-seven hours. While the vehicle's attempts to return to the surface didn't succeed, they managed to bump it along the bottom. As luck would have it, that allowed *Tethys* essentially to plow its way through Monterey Bay, eventually arriving at the shoreline in the vicinity of Santa Cruz in the wee hours of

Saturday morning. When *Tethys* finally reached the surface, its satellite system started sending reports again, and everyone on the team felt their cell phones vibrate in their pocket or ping from their nightstands. We shook off sleep to converge on the vehicle's location, and as we did, we realized one of our team was just a few hundred yards up the beach, having a party with his family. It was the only vehicle recovery I can recall that involved beer.

Imagine, for a moment, if this same situation had happened beneath the ice on Europa. In this scenario, the autonomous vehicle we've sent off into the black has no human minders that can come looking for it. If it gets stuck on the bottom of this alien world, with nine miles of ice and 390 million miles of space between it and Earth, there's no chance of a fortuitous recovery party; there's no coming back from such a mishap. An expensive and irreplaceable robot, and a scientifically indispensable mission, would simply be lost.

We don't yet have the technology for an autonomous vehicle to recover from such a disastrous failure. But we are starting to get closer than you might expect. The data from *Tethys*'s 2013 sinking in Monterey was actually one of the examples that fed into a research program I started to address the problem, initiating it at MBARI and bringing it to WHOI when I moved in 2014. The challenge was to design systems that could detect failures unanticipated by the designers or operators. The classic machine-learning approaches revolve around learning what doesn't work by detecting failures from past experiences. Instead, we specifically wanted to teach the robot to detect something that had never happened before and that we may not even know how to describe.

"At the end of the day, the goal is an unsupervised understanding of the world," says Yogi Girdhar, a WHOI computer scientist who was one of my collaborators on this project. Perhaps counterintuitively, he says, "you don't want to train the AUV about the things that are *not* okay, because the things that are not okay are extremely rare, so there's no way to come up with a dataset to train it with."

Ben Yair Ranaan, then at MBARI, was the one who figured it out: The computer's algorithm was told to continuously try to describe what it was doing based on its prior experiences. If it was not able to describe what was happening, that suggested something was going wrong.[4] This system takes the sort of error recognition that will be used on *Europa Clipper* and pushes it one step further, allowing it to recognize problems that fall outside of those that can be modeled based on past experiences.

That algorithm is now used in *Tethys* systems to help it decide when it needs to call for help. In the future, Girdhar says, it could be combined with other new forms of machine learning to actually adapt to the problem at hand. This type of machine learning, known as reinforcement-based learning, is modeled on how the human brain learns in real life: The computer is allowed to try to complete a set task using trial and error, evaluating how close each attempt gets to the "rewarded" goal. This type of algorithm could allow a vehicle to relearn how to swim if it lost a propeller, or how to complete a scientific mission if one of its instruments went down and it had to rely on a backup.

At WHOI, Girdhar uses similar forms of machine learning to work on another goal that could enable exploration in our world and beyond. His focus is on how to program "curiosity" into exploration robots, in order to give them the ability to discover phenomena that might not have been expected.

"Often you don't know what you're looking for until you're actually looking at it," he explains. "In all of the ocean worlds, you don't have enough prior data that you know what you're looking for. We need robots that can operate in these worlds efficiently, and the question is: Can they identify what is interesting and what should they be paying attention to?"

When most robots are deployed, they're performing relatively simple behaviors on a pre-planned trajectory. But if you want a robot to look for features that are sparsely distributed or wholly unexpected, this sort of programming doesn't work. Imagine, for example, sending one of these classically programmed AUVs to complete one-mile transects below the ice on Europa, taking water samples and photographs of the ice-ocean margin at every meter. The robot would get a comprehensive view of what this margin looked like across that one mile, but the odds are that much of that span would contain little to no data. But now imagine that, in one of those frames, the camera caught a tantalizing glimpse of something bioluminescent clinging to the underside of the ice. (As you might recall, such a brief glimpse is exactly how scientists first discovered hydrothermal vents in the Galápagos.) After reviewing the photographs, the scientists back on Earth could send that robot back to investigate—but there's a chance that whatever it spotted might no longer be there.

Girdhar's programming instead teaches robots to go after what's surprising, after what they don't understand. "The robot is continuously, without annotations from a human, learning a high-level representation of the world, and each image it captures, it incorporates into its model," he explains. "It knows what type of environment it's in and can ask: Given the context of where I am and what I've seen, which part of this image doesn't make sense? And if the robot doesn't understand it, it goes in the direction where it's most confused."

The first AUV created to follow one such program, launched on a coral reef, behaved a lot like a curious puppy, largely ignoring the blank sandy areas and lingering over the spiky, textured shapes of coral heads. In our hypothetical example, such a program operating on an AUV on Europa would immediately recognize that a glowing feature on the underside of the ice was a new and strange thing that didn't fit with the rest of the environment, and go to investigate.

Over the years, the program developed in Girdhar's lab has been refined as the "brain" of CUREE, the Curious Robot for Ecosystem Exploration. CUREE was made to explore coral reefs, which are so complex that surveys are usually still done by human divers; by giving CUREE the ability to make decisions about how to direct its own behavior, using a combination of visual, audio, and mapping data, the robot has already shown that it can perform complicated tasks in a complicated environment, such as identifying the preferred habitat of a specific species of shrimp.

Girdhar's students are also applying versions of this machine learning to teach AUVs to evaluate which parts of their environment to pay attention to. This lets the robot choose to ignore things that aren't essential—another behavior that would be valuable on a distant ocean world, where data storage and communications will be limited, so missions could avoid clogging either with lots of unneeded data.

* * *

To understand the potential impact of this type of artificial intelligence on the world of autonomous exploration, it helps to understand how far it has come. Not so long ago, artificial intelligence was a phrase to be avoided in serious engineering development projects. This was certainly the case when I started the AUV laboratory at MIT. There had been cycles of hype, first in the 1960s, and then in the 1980s, as early forms of artificial intelligence came out of laboratories and proponents promised to revolutionize computing and create machine intelligence. Those periods of excitement and investment were followed by long winters as expectations were dashed.

Over the past two decades, however, several long-awaited developments intersected, leading to tremendous advances. As powerful computers became more common, enormous amounts of digital data became available, which could be used to train machine-learning programs. Computation power increased rapidly in the form of graphical processing units—originally designed to render 3D graphics but can also solve other difficult problems incredibly quickly. Machine learning matured significantly with the introduction of neural networks, computer systems modeled on the human brain, which use different layers of programming to process data, identify what's important, and work together to decide what to do next.

Combined, these advances have revolutionized seemingly intractable problems like computer vision, speech recognition, and most recently, creativity. They take manifest forms, ranging from very sophisticated chatbots (such as ChatGPT), to image and video generation, to a host of specialized systems providing better-than-human performance in tasks ranging from image recognition to protein folding.

The challenge from here is building a foundation that will allow engineers to design with AI confidently. We would like to send our AUV under the Arctic ice shelf, secure in the knowledge that our AI systems are not going to fail after some un-anticipated interactions with the environment (like a shark attack!) and jeopardize the system. Indeed, we would like our AUV's AI to be sophisticated enough that it could detect, diagnose, and work around inevitable failures, improving system reliability rather than potentially compromising it. Although the world of AI is moving quickly, we are not yet close to that goal.

The issue comes from the fact that these artificial intelligence systems do not emerge from a "theory of intelligence"—in other words, rather than being deliberately designed around central theories, many recent advances have come from experimentation. As a result, even the designers of new AI are uncertain about why their systems work as well as they do. This is a problem, because without that understanding, it's hard to predict when these systems will fail.

This unknown differs drastically from the rest of engineering. The highly complex devices in the world around us, from the power systems that carry electricity across continents to the aircraft that fly us around the world, exist because we can guarantee their performance. When engineers design a bridge, they apply their understanding of materials, structures, and failure modes to ensure that they understand the limitations of their design. When failures occur—for example, if a bridge collapses—the failure is studied exhaustively and fixes found. That failure becomes yet another case study that future designers will draw on in ensuring that new bridges are even safer.

Ultimately, the success of engineering rests on a strong theoretical foundation, but it also critically depends on society as a whole: high-integrity engineering processes, ethical standards, and strong institutions (including universities) that teach and require these processes and standards—as well as a rational legal framework and thoughtful governance and policy to hold people accountable.

Many of these activities that guarantee we can trust the engineering around us can slow progress. In our society, there has always been tension between the fast-moving innovators and the conservative framework of assurance. That tension is particularly visible today in AI, as companies rush generative AI systems from the laboratory to deployment. We talk about "benchmarks" to evaluate system performance, and "guardrails" to catch AI when it begins to fail, but the discipline of AI assurance has only a few weak tools in its toolbox.

In part because of this lag in reliability, despite their potential, it will take some time before the kind of adaptable AI we're testing in our own ocean ends up exploring oceans on other worlds. Both Kampmeier and Doyle emphasized that NASA tends to be conservative when it comes to new technology: "NASA is intentionally a little bit behind," Kampmeier says. "On these big flagship missions [like *Europa Clipper*], we have one shot to get this spacecraft up, and it has to work. So one of our driving concerns is reliability: Is it going to work when it needs to work?"

Doyle also pointed out that currently none of the robots that NASA sends into space are equipped with machine learning. Instead, all of that processing is done on the ground back on Earth, and then the models that they create are sent to the vehicle's computer to execute. This creates an inevitable lag for a vehicle trying to respond to its environment. But just as the science objective shifted opinions about AI for the Mars 2020 landing, the possibility of limited communications beneath the ice of an ocean world could certainly push the agency toward installing machine learning aboard one of their probes.

Fortunately, there are strong ties between the space and marine robotics communities, which will continue to help advance and test such systems to the benefit of both. At one point during my time at MBARI, my engineering team held a retreat with the NASA Jet Propulsion Laboratory (JPL) at Big Sur for several days of comparing notes, discussion, and exploration of how we might work together. (The JPL folks also brought a satellite communication system along so we could watch the playoffs, though I've forgotten which sport or who was on the field.)

My take-home from those meetings was that we had a lot going for us in the ocean. Our inability to communicate through water with radio forced us to make our systems autonomous. In contrast, space missions tend to have excellent communications, so while autonomy is an essential capability for an ocean marine robot, it is a "nice-to-have" for a lot of spacecraft— and a nice-to-have that might fail. Further, compared with a space mission, testing robots in the ocean is less costly, and we tend to get our hardware back. Our space colleagues were struggling to get their code on spacecraft and envied our ability to "just take our stuff to sea." There was one area in which they were indisputably supreme, however: NASA budgets are to die for.

Now that I'm at Johns Hopkins, I'm getting to revisit those days in style. One of the best-kept secrets is the growing prominence of Baltimore in space science and industry. The Hubble and Webb space telescopes are operated from a building a short walk from my faculty office. Johns Hopkins is devising and operating a progression of ever more sophisticated robotic space missions. Space companies are also relocating to Baltimore: Rocket Lab, a company that makes lightweight rockets to launch small satellites, opened a manufacturing facility here late in 2023.

The larger trends are, if anything, even more exciting. Ocean world missions aren't a distant aspiration: The *Dragonfly* mission is about to go into its build phase, designed by engineers in Johns Hopkins' Building 200, visible across the green space from my desk at the Applied Physics Lab. Humans are returning to the moon, this time (we hope) to stay. Gateways to orbit are multiplying: Costs are dropping, and new low-cost CubeSats can be built and operated by college students. We're starting a space engineering degree in mechanical engineering and are in the process of hiring faculty who will help pioneer the educational and research frontier of the growing space robotics industry. These are incredibly exciting days for those of us exploring space and the ocean alike—I just wish I were twenty years younger.

I've been humbled many times by the ocean. More than once my team has returned with a long list of lessons learned gained through sleepless hours in cold and rough seas. But over the span of years we've gone from early simple robots with a few sensors that would only last an hour or two beneath the waves to vehicles that will run for weeks carrying genomic laboratories. My enthusiasms are shifting from my first love, which is to create the new technologies, the new robot, the new observation capabilities, to planning for enduring observation campaigns—for example, to understand methane venting in the Arctic.

When I was younger and dreamed of exploration, I never dreamed of robots, yet robots have taken me to the far reaches of the planet. Today robots are moving to the center of the exploration enterprise—they've become a bridge between human curiosity and environments we cannot directly reach. We're creating machines that extend our senses, that respond to fleeting phenomena we might otherwise miss and that deepen our understanding of these remote frontiers. For all my excitement about the pace of development and the new capabilities we're building, the real revolution is only beginning. It's rooted in how these machines will transform our understanding of our environments, allowing us to balance discovery with steward-ship, supporting a thriving human society while safeguarding the ecosystems we depend upon. The more they extend our reach, the more they challenge us to redefine what it means to coexist with the world around us.

Epilogue

IN THE EARLY DAYS of my marine robotics career, I was intensely focused on individual robots. We deployed them from a ship, and they operated through their journeys with little or no supervision. They had to be self-sufficient: carrying their own power, navigating with their onboard sensors, detecting and responding to potential threats, and, of course, executing their mission without asking humans back on shore for instructions.

The future looming before us is likely a very different place. Our robots will work in teams, with many small vehicles accomplishing surveys far faster and using less energy than a single large vehicle. Robotic surface vehicles will support undersea robots by providing communication relays, navigation aids, and solar panels that recharge a diving robot at the surface. They might even provide simple servicing capabilities or tow a broken vehicle back to shore.

We're only beginning to develop these types of sophisticated multiplatform systems, but they're already attracting investment on the order of billions annually in the spheres of energy, scientific research, defense, and marine shipping. A company called Ocean Infinity now operates multiple AUVs for any given mission. These missions are supported by a single crewed surface vessel, but Ocean Infinity is working heavily on replacing that crewed surface vessel with a fleet of autonomous surface vessels, which will support its mirror of an underwater fleet.

As we continue to improve our robots, they will be less dependent on ships, gradually shifting the calculus from minimizing costly ship time by limiting time at sea, to maximizing robotic output by keeping our AUVs at sea. These days, I think most about the potential of such systems in the Arctic, where climate change is happening the fastest, with dramatic effects on its ecosystems. For example, large methane deposits in shallow Arctic sediments are being released as surface waters warm, resulting in methane bubbling from the seafloor. Methane is a strong greenhouse gas, prompting us to ask: How much methane is in the Arctic, and how much is being released? Is it being consumed by microbes, or is it making it to the atmosphere? A grand challenge of Arctic oceanography, to my mind, would be to establish a network of robots providing a year-round pervasive presence in the Arctic Ocean to answer this and many other questions.

At the very least, an ocean full of robots could change the economics of monitoring and mapping that we're currently lacking in so much of the ocean, but especially in places like the Arctic. Yet my hope is that these systems could be used broadly and address the very problems that we've discussed.

Technology is already taking huge leaps ahead to make an industrial ocean more sustainable than previous extractive industries. But providing enough food, power, and minerals to support our growing planet will require walking a delicate tightrope. Specialized, autonomous robotic systems could provide the support and oversight needed to help make ocean-based industries better than their predecessors: repairing nets on ocean-based farms and monitoring the health of farmed fish, conveying power to and from underwater assets like wave-energy machines, and monitoring deep-sea mining activities with cameras and water-column sensors, ensuring that industrial activities in the deep sea are carried out responsibly.

But it's also worth considering that these advancements could make our seas a flash point. The ocean has always been valuable but largely as a connector between regions of commerce and regions with resources. Wars at sea have been fought to project power onto land. That's all changing, as a nation's industrial heart no longer stops at the ocean edge. Industry increasingly continues out to the coastal environment, into wind farms, oil and gas platforms, and aquaculture. If deep-sea mining proceeds, it will extend to the deepest parts of the seafloor, both within national borders and in parts of the ocean that are supposed to be shared for the benefit of all.

As a consequence, the ocean is becoming a point of contention between great nations: See, for example, China's history of increasingly aggressive claims over the South China Sea, including over portions of their neighbors' exclusive economic zones. International diplomatic efforts have made an effort to forestall conflict, but rulings from the Court of Arbitration in the Hague have been rejected by China. Treaties and agreements are only effective if the countries in question choose to follow them. As China moves to mining activities in global waters, it is not unreasonable to ask whether this might lead to conflict on an even larger stage. Recent political changes in the US have made this future even more uncertain.

It's not a stretch to envision that robots could play a role in this tension, too. The technology needed for a bad actor to sabotage an oil well, cut a communication cable, or disrupt a deep-sea mining operation is becoming more and more accessible. Autonomous underwater vehicles once stretched the state of the art in navigation, computation, sensing, and communication; today, we could replace their many tens of thousands of dollars of hardware with a single 2020s-era cell phone. The software to run those vehicles, once the province of a few specialized laboratories, can now be downloaded from the web. These are wonderful developments, as they democratize ocean access. But as we've learned, malevolent actors can take our tools and turn them against us.

How will nations navigate their self-interests and the broader challenges that face humanity? The answers to those questions will determine the world our grandchildren inherit. These advancements can provide us with a safer and more stable world, and as we perfect this technology at home and wrangle with its social implications, we're preparing it to head out on its next grand adventure in the universe. By the time our children and grandchildren are in that world, we may be venturing to space in large numbers and have more specialized robotic systems to augment our capabilities in the outer solar system. They will work together to harvest materials from rocky planets and asteroids to repair each other and themselves, tag-team to send messages back home, and work in groups to search ice-capped oceans and methane seas for the first signals that we are not alone in our solar system. Perhaps one day they might even build homes, providing a safe landing pad for humans as we take our first steps on other worlds. Many of the lessons enabling our space robot cousins will be learned in Earth's ocean.

We are entering a new phase of the human relationship with the ocean, but as we do so; the ocean is changing as well. We now know that the ocean is not limitless, capable of accepting our waste while providing oxygen, food, and a comfortable climate free of charge. One of the first and most important contributions of our new robotic technologies is to remove the veil obscuring the internal workings of the ocean and human activities in that domain. You can't protect what you don't understand, and international agreements are useless without the tools for verification.

While this is a book about technology, perhaps the most important conclusion to take away is that for all the benefits we hope to gain, technology also creates choices. How technology shapes our future oceans will now depend on how carefully we implement it. Having the technology to become a sea-supported society is not sufficient: We also need to conduct the hard practical and ethical conversations about how that technology will be used and what the impacts may be. Will that result in a better home planet and a better understanding of our place within the universe? That is up to all of us to determine.

Acknowledgments

JB: ANY READER OF THIS BOOK will surely realize that developing and fielding ocean robots requires a team willing to commit a part of their life to extremely demanding conditions. I have been truly blessed to work with wonderful and talented individuals throughout my career. In the writing of this book, I was cautioned early on that I was adding too many people to the story, and as a consequence, I have found myself leaving out mentions of individuals and interactions that most profoundly shaped me as a researcher. Here in the acknowledgments, I hope to correct some of those omissions in a small way.

As I write this, I'm just weeks from having celebrated my wife's and my fortieth anniversary. I'm fortunate to be married to someone who is smarter, harder-working, and a better person than I am and who is simultaneously my most effective critic and strongest supporter. Debbie, and our two daughters, Sarah and Elizabeth, put up both with my extensive absences and with the second-order effects of the tremendous stress and pressure I felt, particularly in the early AUV era. During my MIT days, my neighbors joked that she had a virtual husband. I owe an enormous debt of gratitude to my family for sticking with me through what was at times some heavy weather.

The Sea Grant team that spanned from the earliest days though my departure from MIT included many exceptional individuals. Tom Consi, Jim Bales, and John Leonard were the core team that committed their ingenuity and energy to our shared vision of an ocean full of robots. They've all gone on to their independent successes, but I have vivid memories of each of them at sea, in ice camps, and during late nights in the lab as we worked to bring our creations to life. We all committed our early careers to the development of these systems when—to an impartial observer—it was far from clear that they would be useful to anyone before we hit retirement age! I owe a special debt to Henrik Schmidt, who introduced me to the Arctic and opened my eyes (if you can pardon the mixed metaphor) to the acoustic ocean. Finally, Chrys Chryssostomidis, the director of Sea Grant, deserves a special place in marine technology history for his role in nurturing, mentoring, and shaping not just AUVs, but a host of key undersea technologies.

Frank van Mierlo, my partner in founding Bluefin, was the business driver and engine behind making our early AUV company a success. He took the helm of Bluefin at a time when our prospective customers were uncertain of the value of a marine robot, our vehicles were barely out of the laboratory, and investors would not answer our calls. Through Frank and Bluefin I had my first exposure to decision processes in the energy industry. Today, the marine robotics industry includes a growing number of companies started by individuals who started their AUV life at Bluefin.

MBARI is the place where I matured from a researcher to a leader of complex ocean technology projects. Marcia McNutt recruited me, and while working for her I learned people skills that I didn't know I needed, learned to think hard about the problems I committed to, and perhaps most importantly, saw that a healthy family life was achievable even while pushing the limits of the possible. At MBARI, Brett Hobson and Bill Kirkwood became great partners, each teaching me important lessons about leading engineering teams and succeeding with complex technology at sea. Together we built operational capabilities that are second to none. My close friendship and partnership with Yanwu Zhang was instrumental in helping turn vague questions about performance and economics into analytical problems with answers that in turn shaped how we designed our observation systems and our vehicles. My science colleagues, including Ed Delong, Chris Scholin, Ken Johnson, and Francisco Chavez challenged the scientific-value proposition of marine robotics, ultimately greatly sharpening our focus and increasing our impact.

Colleagues at Woods Hole Oceanographic Institution have touched every part of my marine robotics career. Al Bradley mentored me, and then my students, on the unwritten secrets of making complex technology work in the ocean. Dan Fornari pulled me into the Deep Submergence Advisory Committee, which provided a front-row seat—and even a small role—in the transformation of a world-class operational facility from ALVIN to its current portfolio of ALVIN, ROVs, AUVs, and more. Dana Yoerger has been a partner in many AUV and seagoing projects, sometimes started by me, and sometimes started by him.

My education in the technological challenges and opportunities for defense came through service on the Naval Research Advisory Committee and the Secretary of the Navy Advisory Board. Through those organizations I met and worked with a diverse and profoundly committed group of individuals dedicated to ensuring the naval enterprise would be prepared to overcome the tremendous challenges the future threatens. Chief among these was my good friend and mentor John Sommerer who preceded me as NRAC Chair. I owe profound thanks to my NRAC and SNAP colleagues for their mentorship and comradery during those ten wonderful years.

I have found a new home at Johns Hopkins University, in an organization much broader in intellectual scope and engagement than an oceanographic organization. In our new enterprise focused on the assurance of autonomous and AI systems, my Hopkins colleagues live on the Homewood campus, at the Applied Physics Laboratory, on the medical campuses, and in Washington, DC. Once again, I am at the bottom of a steep multidisciplinary learning curve! I would like to thank my new colleagues for welcoming me, and engaging in the thoughtful and challenging discussions that are the hallmark of a truly world-class research enterprise. A particular shout-out to the Penny Black Faculty Salon community members, who have come to me to typify the deep spiritual reward of intellectual engagement across disciplinary boundaries.

This book would not exist without the fabulous people at the Johns Hopkins University Press, the Office of Research, and Bloomberg Philanthropies. Elements of this have been bouncing around in my head for more than two decades, but it was not until I came to Johns Hopkins as a Bloomberg Distinguished Professor that all the right pieces fell into place. Director of Strategic Engagement Anna Marlis Burgard—our editor, creative goad, and taskmaster—shepherded me through the early stages and then introduced me to Claudia Geib. Claudia could not have been a more articulate, knowledgeable, and gracious colleague for coauthoring this work. Thank you, Claudia, for your partnership in going from concept to reality in the creation of this book.

* * *

CG: THIS BOOK WOULD NOT HAVE HAPPENED without the veritable Greek chorus of voices who helped to shape its ideas, who lent their voices in the form of quotes, who answered my many questions and corrected my misconceptions. Thank you to everyone who I interviewed: Diva Amon, Robert Braun,

Sallie Chisholm, Tom Curtin, Ed DeLong, Richard Doyle, Dan Fornari, Halley Froehlich, Yogi Girdhar, Julie Huber, Christian Katlein, Jenny Kampmeier, David Kelly, Marty Klein, Michael Lawson, Andrew Lipsky, Marcia McNutt, David Mindell, Alison Murray, George Nardi, Sarah Lester, Louise Prockter, Chris Scholin, Andrew Thaler, Bob Thresher, and Morgan Trexler. Thanks, too, to all of the communications folks at Johns Hopkins' Applied Physics Laboratory, the Jet Propulsion Laboratory, and NASA Central for making many of these interviews happen.

I owe a great deal of thanks to Jim, whose time, expertise, and shared love of science fiction made for a fascinating and fun co-authorship; to Johns Hopkins' Director of Strategic Engagement, Anna Marlis Burgard, and Johns Hopkins University Press's Executive Editor, Matthew McAdam, for guiding the book with their suggestions, refinements, and assistance in structural quandaries; and to Michael Greshko, for connecting me with the Johns Hopkins team in the first place.

Finally, my gratitude as always to Matt: for his love, unconditional support, listening ear, and occasional help with questions about how engineering works.

Notes

PREFACE

1. James G. Bellingham, Clifford A. Goudey, Thomas R. Consi, James W. Bales, Donald K. Atwood, John J. Leonard, and Chryssostomos Chryssostomidis, "A Second Generation Survey AUV," *Proceedings of IEEE Symposium on Autonomous Underwater Vehicle Technology,* 1994, 148–55. https://doi.org/10.1109/AUV.1994.518619.

CHAPTER 1. THE CHALLENGE OF SEARCHING THE SEAS

1. Mark Kurlansky, *Cod* (Knopf Canada, 1997).

2. David Abulafia, *The Boundless Sea* (Oxford University Press, 2019).

3. Sarah Gibbens, "'Sea Nomads' Are First Known Humans Genetically Adapted to Diving," *National Geographic,* April 19, 2018, https://www.nationalgeographic .com/science/article/bajau-sea-nomads-free-diving-spleen-science.

4. Armand Marie Leroi, *The Lagoon: How Aristotle Invented Science* (Penguin, 2014).

5. Throughout this chapter, we have chosen to use the word "discover" when referring to Europeans' first knowledge of many places around the globe. This usage is certainly Eurocentric, as much of modern history is: Columbus did not "discover" the Americas, as there were untold Indigenous civilizations already there, just as other European explorers did not discover peopled places across the Arctic and Polynesia. When I use the word, it is for the sake of brevity or lack of a good alternative and with the recognition of the colonizing history of the term.

6. Philip Edwards, ed., *The Journals of Captain Cook* (Penguin, 1999).

7.   The Darwin Correspondence Project: *Barnacles,* accessed December 9, 2024, https://www.darwinproject.ac.uk/barnacles.

8.   Helen Scales, *The Brilliant Abyss* (Atlantic Monthly Press, 2021).

9.   Thomas R. Anderson and Tony Rice, "Deserts on the Sea Floor: Edward Forbes and His Azoic Hypothesis for a Lifeless Deep Ocean," *Endeavour* 30, no. 4 (2006): 131–37, https://doi.org/10.1016/j.endeavour.2006.10.003.

10.   Anderson and Rice, "Deserts on the Sea Floor."

11.   Graham Bell, *Full Fathom 5000: The Expedition of HMS Challenger and the Strange Animals It Found in the Deep Sea* (Oxford University Press, 2022).

12.   "The temperature data launched an entire discipline called physical oceanography. It's our first picture of the physical ocean all around the globe, and it remains a really important touchstone for looking back at the history of the climate," says Josh Willis, a climate scientist at NASA who has used temperature data from *Challenger* in his research on global climate trends. Kate Golembiewski, "H.M.S. Challenger's First Real Glimpse of the Deep Oceans," *Discover Magazine,* April 19, 2019, https://www.discovermagazine.com/planet-earth/hms-challenger-humanitys-first-real-glimpse-of-the-deep-oceans.

13.   The Leonardo Museum: Codex Atlanticus, folio 881r, *Submarine,* accessed December 9, 2024, https://www.leonardo3.net/en/l3-works/machines/1467-mechanical-submarine.html.

14.   Naval History and Heritage Command, USS Holland (Submarine Torpedo Boat #1), accessed December 9, 2024, https://www.history.navy.mil/our-collections/photography/us-navy-ships/alphabetical-listing/h/uss-holland--submarine-torpedo-boat--1-0.html.

15.   Gary Weir, *An Ocean in Common* (Texas A&M University Press, 2001).

16.   G. Pascal Zachary, *Endless Frontier: Vannevar Bush, Engineer of the American Century* (Free Press, 1997).

17.   Angela D'Amico and Richard Pittenger, "A Brief History of Active Sonar," *Aquatic Mammals* 35, no. 4 (2009): 426–34, https://doi.org/10.1578/AM.35.4.2009.426.

18. Weir, *An Ocean in Common.*

19. Casey MacLean, "World War I on the Homefront," *NOAA National Marine Sanctuaries,* May 2018, https://sanctuaries.noaa.gov/news/may18/world-war-i-on-the-homefront.html.

20. Li Zhou, "This Map Shows the Full Extent of the Devastation Wrought by U-Boats in World War I," *Smithsonian Magazine,* May 7, 2015, https://www.smithsonianmag.com/history/map-shows-full-extent-devastation-wrought-uboats-world-war-i-180955191.

21. Catherine Musemeche, "Mary Sears' Pioneering Ocean Research Saved Countless Lives in WWII," *Smithsonian Magazine,* July/August 2022, https://www.smithsonianmag.com/history/mary-sears-pioneering-ocean-research-saved-countless-lives-wwii-180980325/.

22. Weir, *An Ocean in Common.*

23. Naomi Oreskes, *Science on a Mission: How Military Funding Shaped What We Do and Don't Know About the Ocean* (University of Chicago Press, 2021).

24. Oreskes, *Science on a Mission.*

25. Ed Offley, "U-Boats in the Gulf," *USA Today,* https://stories.usatodaynetwork.com/uboatsinthegulf/.

26. Mark Carlson, "The U.S. Navy's Defective Mark 14 Torpedo," *Warfare History Network,* June 2017, https://warfarehistorynetwork.com/article/the-defective-mark-14-torpedo/.

27. National Park Service and the Arizona Memorial Museum Association, "The Silent Services: Submarines in the Pacific," *A Guide to War in the Pacific,* https://www.nps.gov/parkhistory/online_books/wapa/extContent/wapa/guides/offensive/sec6.htm.

CHAPTER 2. OPENING THE DOOR

1. Robert F. Thompson, interview by Kevin M. Rusnak, August 29, 2000, NASA Johnson Space Center Oral History Project https://historycollection.jsc.nasa.

gov/JSCHistoryPortal/history/oral_histories/ThompsonRF/Thomp-sonRF_8-29-00.htm.

2. James Schefter, *The Race: The Complete True Story of How America Beat Russia to the Moon* (Knopf Doubleday, 2000).

3. University of Rhode Island and Inner Space Center, "History of the SOFAR Channel," Discovery of Sound in the Sea, https://dosits.org/science/movement/sofar-channel/history-of-the-sofar-channel/.

4. It's often stated that SOFAR bombs were used during World War II, but this appears to be a historical misconception; I could find no record of pilots being rescued thanks to SOFAR bombs during the war, and contemporaneous sources suggest that they weren't ready for use during the war. See, for example, "Science: Sofar," *TIME Magazine,* February 4, 1946, https://content.time.com/time/subscriber/article/0,33009,776648,00.html.

5. Norman Polmar, "What Killed the Thresher?," *Naval History Magazine,* April 2023, https://www.usni.org/magazines/naval-history-magazine/2023/april/what-killed-thresher.

6. Naomi Oreskes, *Science on a Mission: How Military Funding Shaped What We Do and Don't Know About the Ocean* (University of Chicago Press, 2021).

7. Will Forman, *The History of American Deep Submersible Operations* (Best Publishing, 1999).

8. William Connelly, "Oceanography: A 'Wet NASA,' Will Nixon Take the Plunge?," *Science* 168, no. 3927 (1970): 98–101, http://doi.org/10.1126/science.168.3927.

9. Most of the information in this chapter, unless otherwise indicated, comes from James G. Bellingham, "Have Robot, Will Travel," *Methods in Oceanography* 10 (2014): 5–20, https://doi.org/10.1016/j.mio.2014.10.001.

10. Max Deffenbaugh, "A Matched Field Processing Approach to Long Range Acoustic Navigation" (master's thesis, Massachusetts Institute of Technology, 1991), https://core.ac.uk/download/pdf/4399321.pdf.

CHAPTER 3. TEAMWORK

1. "Seasat Celebrates Landmark in Remote-Sensing History," NASA Jet Propulsion Laboratory, June 2013, https://www.jpl.nasa.gov/images/pia15816 -seasat-celebrates-landmark-in-remote-sensing-history

2. Allan R. Robinson, Michael A. Spall, Wayne G. Leslie, Leonard J. Walstad, and Dennis J. McGillicuddy, "Gulfcasting: Dynamical Forecast Experiments for Gulf Stream Rings and Meanders, November 1985–June 1986," *Harvard Open Ocean Model Reports: Reports in Meteorology and Oceanography*, no. 22 (1987), https://apps.dtic.mil/sti/citations/ADA221482; Allan R. Robinson, Michael A. Spall, Leonard J. Walstad, and Wayne G. Leslie, "Data Assimilation and Dynamical Interpolation in GULFCAST Experiments," *Dynamics of Atmospheres and Oceans* 13, nos. 3–4 (1989): 301–316, https://doi.org/10.1016 /0377-0265(89)90043-2.

3. *University of Rhode Island: Synoptic Ocean Prediction Experiment (SYNOP)* (undated), https://web.uri.edu/gso/research/dynamics/projects /synop/.

4. Allan R. Robinson, Scott M. Glenn, Michael A. Spall, Leonard J. Walstad, Geraldine M. Gardner, and Wayne G. Leslie, "Forecasting Gulf Stream Meanders and Rings," *Oceanography Report* 70, no. 45 (1989): 1464–1473, https://doi.org/10.1029/89EO00346.

5. Kenneth Chang, "Taking the Oceans' Pulse, with Help from Robot Subs," *New York Times*, September 30, 2003, https://www.nytimes.com /2003/09/30/science/taking-the-oceans-pulse-with-help-from-robot-subs .html.

6. Henry Stommel, "The Slocum Mission," *Oceanography* 2, no. 1 (2015): 22–25, https://doi.org/10.5670/oceanog.1989.26.

7. Steve R. Ramp et al., "Preparing to Predict: The Second Autonomous Ocean Sampling Network (AOSN-II) Experiment in the Monterey Bay," *Deep Sea Research Part II: Topical Studies in Oceanography* 56, nos. 3–5 (2009): 68–86, https://doi.org/10.1016/j.dsr2.2008.08.013.

8. Rebecca Burke, "NAVOCEANO Forges Ahead, Surpassing Unmanned Systems Milestone," *Military News*, July 30, 2018, https://www.militarynews .com/news/navoceano-forges-ahead-surpassing-unmanned-systems -milestone/article_a758eaba-941d-11e8-b75c-1b2f02a11ee0.html.

9. Integrated Ocean Observing System, https://ioos.noaa.gov/data/.

10. National Science Foundation, "The Vision," *Ocean Observatories Initiative*, https://oceanobservatories.org/the-vision/.

11. Australian Transport Safety Board, "The Operational Search for MH370," https://www.atsb.gov.au/publications/investigation_reports/2014/aair /ae-2014-054.

12. *Reuters Graphics*, "The World's Largest Search" (2017), http://fingfx .thomsonreuters.com/gfx/rngs/MH370-SEARCH/0100316R2NR/.

13. Australian Government, "The Data Behind the Search for MH370," https://geoscience-au.maps.arcgis.com/apps/Cascade/index.html?appid= 038a72439bfa4d28b3dde81cc6ff3214; Katherine Kornei, "Seafloor Data from Lost Airliner Search Are Publicly Released," *Eos* 98 (2017), https://doi.org /10.1029/2017EO078307.

14. UNOLS Fleet Improvement Committee, "U.S. Academic Research Fleet Improvement Plan 2019 Update," 2019, https://www.unols.org/sites/ default/files/Fleet_Improvement_Plan_2019_Final_191009.pdf.

CHAPTER 4. THE LIVING OCEAN

1. Elizabeth Pennisi, "Meet the Obscure Microbe That Influences Climate, Ocean Ecosystems, and Perhaps Even Evolution," *Science News*, March 9, 2017, https://www.science.org/content/article/meet-obscure-microbe -influences-climate-ocean-ecosystems-and-perhaps-even-evolution.

2. John Waterbury, "Little Things Matter A Lot," *Oceanus*, March 11, 2005, https://www.whoi.edu/oceanus/feature/little-things-matter-a-lot/.

3. Vivien Marx, "Why the Ocean Virome Matters," *Nature Methods* 19 (2022): 924–927, https://doi.org/10.1038/s41592-022-01567-3.

4. Brett W. Hobson, James G. Bellingham, Brian Kieft, Rob McEwen, Michael Godin, Yanwu Zhang, "Tethys-Class Long Range AUVs—Extending the Endurance of Propeller-Driven Cruising AUVs from Days to Weeks," *IEEE/OES Autonomous Underwater Vehicles (AUV)*, 2012, 1–8, https://doi.org/10.1109/AUV.2012.6380735.

CHAPTER 5. THE OCEAN IN OUR KITCHENS

1. Hannah Ritchie, "The World Now Produces More Seafood from Fish Farms Than Wild Catch," *Our World in Data*, September 13, 2019, https://ourworldindata.org/rise-of-aquaculture.

2. Leslie Harris O'Hanlon, "Deep-Sea Creatures Yield Treasure Trove of Cancer Drugs," *Nature Medicine* 11, no. 698 (2005), https://doi.org/10.1038/nm0705-698b; Harshad Malve, "Exploring the Ocean for New Drug Developments: Marine Pharmacology," *Journal of Pharmarcy and BioAllied Sciences* 8, no. 2 (2016): 83–91, https://doi.org/10.4103/0975-7406.171700; Helen Scales, *The Brilliant Abyss* (Atlantic Monthly Press, 2021).

3. World Bank Group, "Blue Economy," https://www.worldbank.org/en/topic/oceans-fisheries-and-coastal-economies.

4. Nicolas Gruber et al., "The Oceanic Sink for Anthropogenic CO_2 from 1994 to 2007," *Science* 363, no. 6432 (2019): 1193–1199, https://doi.org/10.1126/science.aau5153.

5. UNESCO, "UNESCO Cautions Ocean Risks Losing Its Ability to Absorb Carbon, Exacerbating Global Warming," April 20, 2021, https://www.unesco.org/en/articles/unesco-cautions-ocean-risks-losing-its-ability-absorb-carbon-exacerbating-global-warming.

6. Megumi O. Chikamoto, Pedro DiNezio, and Nicole Lovenduski, "Long-Term Slowdown of Ocean Carbon Uptake by Alkalinity Dynamics," *Geophysical Research Letters* 50, no. 4 (2023), https://doi.org/10.1029/2022GL101954.

7. Bella Isaacs-Thomas, "When It Comes to Sucking Up Carbon Emissions, 'The Ocean Has Been Forgiving'; That Might Not Last," *PBS News Hour*,

March 25, 2022, https://www.pbs.org/newshour/science/the-ocean
-helps-absorb-our-carbon-emissions-we-may-be-pushing-it-too-far.

8. Homi Kharas and Meagan Dooley, "The Evolution of Global Poverty,
1990–2030," Brookings Institution Center for Sustainable Development,
https://www.brookings.edu/wp-content/uploads/2022/02/Evolution-of-global
-poverty.pdf.

9. Erica Geis, "Hawaii's Ancient Aquaculture Revival," *Biographic,* June 12, 2019,
https://www.biographic.com/hawaiis-ancient-aquaculture-revival/.

10. Larry Pynn, "Clam Digging through 3,500 Years of Indigenous History," *Hakai
Magazine,* February 27, 2019, https://hakaimagazine.com/news/clam-digging
-through-3500-years-of-indigenous-history/.

11. Felix Richter, "Aquaculture Accounts for Half of the World's Fish Supply,"
World Economic Forum, April 29, 2022, https://www.weforum.org/agenda
/2022/11/aquaculture-half-worlds-fish-supply-food-security/.

12. United Nations Climate Change, "Plenty of Fish?," June 10, 2022, https://
unfccc.int/blog/plenty-of-fish.

13. United States Geological Survey, "What A Drag: the Global Impact of Bottom
Trawling," March 14, 2016, https://www.usgs.gov/news/national-news-release
/what-drag-global-impact-bottom-trawling.

14. Karen McVeigh, "Bottom Trawling Releases as Much Carbon as Air Travel,
Landmark Study Finds," *The Guardian,* March 17, 2021, https://www
.theguardian.com/environment/2021/mar/17/trawling-for-fish-releases
-as-much-carbon-as-air-travel-report-finds-climate-crisis.

15. Rebecca L. Lewison et al., "Global Patterns of Marine Mammal, Seabird, and
Sea Turtle Bycatch Reveal Taxa-Specific and Cumulative Megafauna
Hotspots," *PNAS* 111, no. 14 (2014): 5271–5276, http://doi.org/10.1073/pnas
.1318960111.

16. Ray Hilborn et al., "Effective Fisheries Management Instrumental in Improving
Fish Stock Status," *PNAS* 117, no. 4 (2020): 2218–2224, https://doi.org/10.1073
/pnas.1909726116.

17. Carlos Brais Carballeira Braña et al., "Towards Environmental Sustainability in Marine Finfish Aquaculture." *Frontiers in Marine Science* 8 (2021), https://doi.org/10.3389/fmars.2021.666662.

18. Divas Karimanzira et al., "First Testing of an AUV Mission Planning and Guidance System for Water Quality Monitoring and Fish Behavior Observation in Net Cage Fish Farming," *Information Processing in Agriculture* 1, no. 2 (2013): 131–140. https://doi.org/10.1016/j.inpa.2014.12.001

19. Vaggelis Chalkiadakis et al., "Designing a Small-Sized Autonomous Underwater Vehicle Architecture for Regular Periodic Fish-Cage Net Inspection," *IEEE International Conference on Imaging Systems and Techniques (IST)*, Beijing, China, 2017, 1–6, https://doi.org/10.1109/IST.2017.8261525.

20. Lisa Jackson, "Rise of the Machines: Aquaculture's Robotic Revolution," *Global Seafood Alliance Responsible Seafood Advocate*, February 13, 2017, https://www.globalseafood.org/advocate/rise-of-the-machines-aquacultures-robotic-revolution/.

21. SINTEF, "Artifex Project," May 31, 2018, https://www.sintef.no/en/projects/2016/artifex/.

22. Boxfish Robotics. "How We've Solved the Autonomous Underwater Vehicle Puzzle with ARV-I, " April 13, 2023, https://www.boxfish.nz/asset-inspection/autonomous-underwater-vehicle-arv-i/

23. Karimanzira et al., "First Testing of an AUV Mission Planning and Guidance System."

24. MIT Sea Grant, "AUV Lab Vehicles," https://seagrant.mit.edu/auv-lab-vehicles/.

25. Halley E. Froehlich, Rebecca R. Gentry, Michael B. Rust, Dietmar Grimm, and Benjamin S. Halpern, "Public Perceptions of Aquaculture: Evaluating Spatiotemporal Patterns of Sentiment Around the World," *PLoS ONE* 12, no. 1 (2017): e0169281, https://doi.org/10.1371/journal.pone.0169281.

CHAPTER 6. POWERED BY THE OCEAN

1. David Biello, "How Science Stopped BP's Gulf of Mexico Oil Spill," *Scientific American,* April 19, 2011, https://www.scientificamerican.com/article/how -science-stopped-bp-gulf-of-mexico-oil-spill/.

2. Michael Carroll et al., *An Analysis of the Impacts of the Deepwater Horizon Oil Spill on the Gulf of Mexico Seafood Industry*, US Department of the Interior, Bureau of Ocean Energy Management, Gulf of Mexico OCS Region, New Orleans, LA. OCS Study BOEM 2016–020 (2016), https://espis.boem.gov /final%20reports/5518.PDF.

3. Martha Harbison, "More Than One Million Birds Died During Deepwater Horizon Disaster," *Audubon Magazine,* May 6, 2014, https://www.audubon .org/news/more-one-million-birds-died-during-deepwater-horizon -disaster.

4. National Ocean Service, "Deepwater Horizon Oil Spill: Long-Term Effects on Marine Mammals, Sea Turtles." *National Oceanic and Atmospheric Administration*, April 20, 2017, https://oceanservice.noaa.gov/news/apr17/dwh-protected -species.html.

5. Patrick T. Schwing et al., "A Synthesis of Deep Benthic Faunal Impacts and Resilience Following the Deepwater Horizon Oil Spill," *Frontiers in Marine Science* 7 (2020), https://doi.org/10.3389/fmars.2020.560012.

6. Joshua Daley, "Deepwater Horizon Site Is Now a Sticky Wasteland Populated by Sickly Crabs," *Smithsonian Magazine,* September 19, 2019, https://www .smithsonianmag.com/smart-news/deepwater-horizon-site-wasteland -populated-sickly-crabs-180973181/.

7. John M. Broder, "BP Shortcuts Led to Gulf Oil Spill, Report Says," *New York Times,* September 14, 2011, https://www.nytimes.com/2011/09/15/science /earth/15spill.

8. Lisa Friedman, "Ten Years After Deepwater Horizon, U.S. Is Still Vulnerable to Catastrophic Spills," *New York Times*, April 19, 2020, https://www.nytimes .com/2020/04/19/climate/deepwater-horizon-anniversary.html.

9. Baker Hughes, "Rig Count Overview & Summary Count," accessed February 25, 2025, https://rigcount.bakerhughes.com/rig-count-overview.

10. Bruce Beaubouef, "Report: Trump Prepares Wide-Ranging Energy Plan to Boost Gas Exports, Oil Drilling," *Offshore Magazine,* December 9, 2024, https://www.offshore-mag.com/regional-reports/us-gulf-of-mexico/news/55248507/report-trump-prepares-wide-ranging-energy-plan-to-boost-gas-exports-oil-drilling.

11. Gulf of Mexico Research Initiative, accessed February 25, 2025, https://research.gulfresearchinitiative.org/.

12. ARCTIS Knowledge Hub, "Arctic Ports," *AMSA Report 2009,* http://www.arctis-search.com/Arctic+Ports.

13. Department of Homeland Security, Office of Science and Technology, "Snapshot: S&T's Arctic Domain Awareness Center Supports U.S. Coast Guard Response Capabilities," April 7, 2017, https://www.dhs.gov/archive/science-and-technology/news/2017/04/07/snapshot-sts-arctic-domain-awareness-center-supports-uscg-capabilities.

14. Ariana Hurtado, "Operators Ramping Up AUV/ROV Efforts Offshore," *Offshore,* August 9, 2023, https://www.offshore-mag.com/vessels/article/14296152/operators-ramping-up-auv-rov-efforts-offshore.

15. V. Masson-Delmotte et al., "IPCC, 2018: Summary for Policymakers," *Global Warming of 1.5°C. An IPCC Special Report on the Impacts of Global Warming of 1.5°C Above Pre-Industrial Levels and Related Global Greenhouse Gas Emission Pathways, in the Context of Strengthening the Global Response to the Threat of Climate Change, Sustainable Development, and Efforts to Eradicate Poverty* (2018): 3–24, https://doi.org/10.1017/9781009157940.001.

16. WHOI Media Relations, "DOE Funding Will Support WHOI Research to Support Sustainable Development of Offshore Wind," *Woods Hole Oceanographic Institution,* November 19, 2021, https://www.whoi.edu/press-room/news-release/doe-funding-will-support-whoi-research-to-support-sustainable-development-of-offshore-wind/.

17. US Securities and Exchange Commission, "SEC Adopts Rules to Enhance and Standardize Climate-Related Disclosures for Investors," March 6, 2024, https://www.sec.gov/news/press-release/2024-31.

18. Deep Sea Conservation Coalition, "Pivotal Shift at the ISA: Nations Rally for Deep-Sea Mining Moratorium," August 3, 2024, https://deep-sea-conservation .org/pivotal-shift-at-the-isa-nations-rally-for-deep-sea-mining-moratorium/

19. Deep Sea Conservation Coalition, "Momentum for a Moratorium," https:// savethehighseas.org/moratorium_2022/.

20. Helen Scales, *The Brilliant Abyss* (Atlantic Monthly Press, 2021).

21. Diva J. Amon et al., "Climate Change to Drive Increasing Overlap Between Pacific Tuna Fisheries and Emerging Deep-Sea Mining Industry, " *npj Ocean Sustainability* 2, no. 9 (2023), https://doi.org/10.1038/s44183-023-00016-8.

22. Scales, *The Brilliant Abyss.*

23. Daniel O. B. Jones et al., "Biological Responses to Disturbance from Simulated Deep-Sea Polymetallic Nodule Mining," *PLoS ONE* 12, no. 2 (2017), https://doi.org/10.1371/journal.pone.0171750.

CHAPTER 7. OCEANS ACROSS THE SOLAR SYSTEM

1. John B. Corliss, Jack Dymond, Louis I. Gordon, John M. Edmond, Richard P. von Herzen, Robert D. Ballard, Kenneth Green, et al., "Submarine Thermal Springs on the Galapagos Rift," *Science* 203, no. 4385 (1979): 1073–1083, https://doi.org/10.1126/science.203.4385.1073.

2. WHOI Archives, "The Discovery of Hydrothermal Vents: 25th Anniversary," *Woods Hole Oceanographic Institution,* 2002, https://divediscover.whoi.edu /archives/ventcd/.

3. Timothy Oleson, "Benchmarks: February 17, 1977: Hydrothermal Vents Are Discovered," *EARTH,* January 4, 2015, https://www.earthmagazine.org/article /benchmarks-february-17-1977-hydrothermal-vents-are-discovered/; Evan Lubofsky, "The Discovery of Hydrothermal Vents," *Oceanus,* June 11, 2018, https://www.whoi.edu/oceanus/feature/the-discovery-of-hydrothermal-vents/;

National Geographic Education, "Deep Sea Hydrothermal Vents," *National Geographic.* https://education.nationalgeographic.org/resource/deep-sea -hydrothermal-vents/.

4. NASA, "Jupiter's Moons: Ganymede," https://solarsystem.nasa.gov/moons /jupiter-moons/ganymede/in-depth/.

5. NASA, "Saturn's Moons: Enceladus," https://solarsystem.nasa.gov/moons /saturn-moons/enceladus/in-depth/.

6. NASA Jet Propulsion Laboratory, "NASA Cassini Data Reveals Building Block for Life in Enceladus' Ocean," June 14, 2023, https://www.nasa.gov /feature/jpl/nasa-cassini-data-reveals-building-block-for-life-in-enceladus -ocean.

7. Kenneth Chang, "Poison Gas Hints at Potential for Life on an Ocean Moon of Saturn," *New York Times,* December 14, 2023, https://www.nytimes.com /2023/12/14/science/enceladus-moon-cyanide-life-saturn.html.

8. NASA, "Saturn's Moons: Titan," https://solarsystem.nasa.gov/moons/saturn -moons/titan/overview/.

9. Kenneth Chang, "An Ocean May Lurk Inside Saturn's 'Death Star' Moon," *New York Times,* January 21, 2022, https://www.nytimes.com/2022/01/21 /science/mimas-ocean-death-star.html.

10. Theresa Machemer, "More Evidence That Pluto Might Have a Subsurface Ocean," *Smithsonian Magazine,* April 1, 2020, https://www.smithsonianmag.com /smart-news/more-evidence-pluto-might-have-subsurface-ocean-180974550/.

11. NASA, "Exoplanet Catalog," https://exoplanets.nasa.gov/discovery/exoplanet -catalog/.

12. Takeshi Naganuma and Hirohiko Uematsu, "Dive Europa: A Search-for-Life Initiative," *Biological Sciences in Space* 12, no. 2 (1998): 126–130, https://doi .org/10.2187/bss.12.126.

13. Isabelle Gerretsen, "Why Nasa Is Exploring the Deepest Oceans on Earth," *BBC,* January 12, 2022, https://www.bbc.com/future/article/20220111-why -nasa-is-exploring-the-deepest-oceans-on-earth; Shannon Stirone, "NASA Is

Preparing for Future Space Missions by Exploring Underwater Volcanoes Off Hawaii," *Popular Science,* August 27, 2018, https://www.popsci.com/nasa -subsea-hawaii/; N. A. Raineault and J. Flanders, eds., "New Frontiers in Ocean Exploration: The E/V Nautilus, NOAA Ship Okeanos Explorer, and R/V Falkor 2018 field Season," *Oceanography* 32, no. 1 (supplement) (2019), https://doi.org/10.5670/oceanog.2019.supplement.01.

14. Yinon M. Bar-On, Rob Phillips, and Ron Milo, "The Biomass Distribution on Earth," *PNAS* 115, no. 25 (2018): 6506–6511, https://doi.org/10.1073/pnas .1711842115.

15. Marina Koren, "There's Hope for Life on Europa, a Distant Moon," *The Atlantic,* October 6, 2022, https://www.theatlantic.com/science/archive/2022/10/jupiter -moon-europa-ocean-life-forms-nasa-clipper-mission/671671/.

CHAPTER 8. COULD DEEP-SEA ROBOTS DISCOVER LIFE IN DEEP SPACE?

1. For an informative and fun primer on the impacts of bit flips, see the episode of the Radiolab podcast on the topic: https://radiolab.org/podcast/bit-flip.

2. Matt Hamblen, "How Perseverance Landed On Mars with Its Terrain Nav System," *Fierce Electronics,* June 22, 2021, https://www.fierceelectronics.com /electronics/how-perseverance-landed-mars-its-terrain-nav-system; Swati Patel, "Impact Story: Terrain Relative Navigation," *NASA,* September 30, 2022, https://www.nasa.gov/directorates/spacetech/Terrain_Relative_Navigation; Richard Doyle, interview by David Zierler, "Richard Doyle, Data Scientist and AI Space Science Researcher," Caltech Heritage Project, February 23, 2023, https://heritageproject.caltech.edu/interviews-updates/richard-doyle.

3. Patel, "Impact Story."

4. Ben-Yair Raanan, James Bellingham, Yanwu Zhang, Mathieu Kemp, Brian Kieft, Hanumant Singh, and Yogesh Girdhar, "Detection of Unanticipated Faults for Autonomous Underwater Vehicles Using Online Topic Models," *Journal of Field Robotics* 35, no. 5 (2017): 705–716, https://doi.org/10.1002/rob.21771.

Index

JOHNS HOPKINS
WAVELENGTHS

Explores the beneficial roles of fungi, their deadly impacts, and how experts are researching treatments to save lives and our food supplies.

A guided journey through the inner workings of Earth, the cloaked mysteries of other planets in our solar system, and beyond.

Artificial intelligence is part of our daily lives. How can we address its limitations and guide its use for the benefit of communities worldwide?

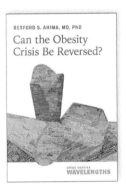

How can we work together to understand the rise of obesity and reverse its related diseases and societal impacts?

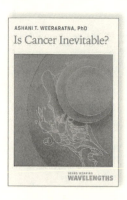

ASHANI T. WEERARATNA, PhD

Is Cancer Inevitable?

WAVELENGTHS

LISA COOPER, MD, MPH

Why Are
Health Disparities
Everyone's Problem?

WAVELENGTHS

How can new understandings
about cancer cell interactions
help doctors better control, and
eventually cure, cancer?

How can we all work together to
eliminate the avoidable injustices
that plague our health care
system and society?

JESSICA FANZO, PhD

Can Fixing Dinner
Fix the Planet?

WAVELENGTHS

How can consumers, nations, and
international organizations work together to
improve food systems before our planet loses
its ability to sustain itself and its people?

 JOHNS HOPKINS
UNIVERSITY PRESS

PRESS.JHU.EDU